Marline-Spike Seamanship

MARLINE-SPIKE SEAMANSHIP

The Art of Handling, Splicing and Knotting Wire

"*The efficiency of a splice is judged on its non-drawing power*"

BY
LEONARD POPPLE

GLASGOW
BROWN, SON & FERGUSON, LTD., Nautical Publishers
4-10 DARNLEY STREET

Copyright in all countries signatory to the Berne Convention
All rights reserved

First Edition	1946
Reprinted	1958
Reprinted	1970
Reprinted	1981
Revised Edition	1994

ISBN 0 85174 621 7
ISBN 0 85174 138 X (First Edition)

© 1994 BROWN, SON & FERGUSON, LTD., GLASGOW, G41 2SD
Made and Printed in Great Britain

FOREWORD

SOME splendid books have been written on the ancient art of wire and rope working, the authors mostly deploring that the art is fast nearing extinction. It was after reading Scot Skirving's and Chas. L. Spencer's fine books on the subject that I decided to attempt to add to their works. If I succeed I shall have reached the pinnacle of my ambition as a rigger, for it had long been said that riggers on the whole are inarticulate. It is my intention to describe and sketch only wire and spike seamanship — work on which, to the best of my knowledge, has never before been published.

This book is written not for the professional rigger, but for the beginner who has no means of learning the fundamentals. There are at least half a dozen methods of turning in the "Eye-Splice" — with the lay of the wire, and against it, with and without the locking tuck, or a mixture of with and against the lay, the Liverpool or spiral method, the merits of which need not be discussed here; for when a simple method has been mastered, the more complicated methods should be soon accomplished. I have yet to see the failure of an Eye Splice though I have seen many wires part below the splice due to the great stress involved. I remember watching a Chinese seaman fitting an eye. His method was, tuck all the strands with the lay, in the "under one and over one style", and on my conferring with him, he assured me that his splice was "All can do". And he was no doubt correct in saying that his splice was as good as any other, for the strain needed to draw an eye must be terrific, if the tucks are equally taut and properly home.

It is the little things that matter in wire splicing, such as efficient whippings and seizings. Some dip the ends of the strands in solder, others prefer twisting the ends up with pliers. I am referring to small wires now, but big or small, the only method worth talking about is the good old common whipping. It stays put if properly executed. Some have a flair for knotting and splicing. To those I would say, pass the good work on, for as long as we have ships, we shall need wire and rope to rig and berth them and men to splice.

It has often been argued, that there will always be plenty to do such work, but there is splicing, and splicing, and the competent rigger is as proud of his skill with the spike, as the academical painter is with his brush, and he has every right to be, for it calls for keen observation and perseverance. My final remarks are, aim at a good style as well as efficiency. Nothing looks worse than the crown of an eye which has lost its virgin lay, a lumpy irregular splice, a loose thimble or a faulty serving. Remembering always that practical rigging is partly a craft, partly a science, which no one dare lay claim to having completely mastered; I find myself thinking, what a lot I might learn from a book on this subject, that someone else might care to write.

Nomenclature — My system of naming no doubt differs from other writers on the subject, but it is true to say that in various parts of Britain the naming of this and that differs. The riggers have their own pet names. It matters little, as far as I can see, whether one terms it the Main heart or the Europe heart, the toe of the thimble instead of the heel, a grummet instead of a grommet or a strop instead of a strap.

CONTENTS

	PAGE
Tools	1
How to Use the Marline-spike	3
A Few Words on Knotting	4
How to Serve	5
How to Worm	7
How to Parcel	7
On Whippings	8
How to Make up a Hank of Spunyarn for Serving	9
How to Join A Broken Serving	9
How to Make up Serving Wire	10
Hints on Splicing in the under and over style	11
How to Splice extra Special Wire in the under and over style	11
Method of Preparing Wire for Splicing in the under and over style	13
The Hawser or Towing Eye and Attenuation	16
The Spiral or Liverpool Styled Eye-Splice	18
How to Splice in the Spiral Fashion	20
The Liverpool or Spiral Short Splice	24
The Short Splice	24
The Reduced Eye Type "A"	26
The Reduced Eye Type "B"	28
The Flemish Eye in Wire	28
The Cut Splice in Wire and Splicing Wire to Rope	30
The Long Splice	31
Wire Seizings	35
On Span Making	37
On Seizing in Small Thimbles	39
The Cropper or Gag Method of Securing Large Thimbles	40
On Stropping Blocks	42
To Seize a Cleat on Davit or Stanchion	44
Wire Net Making	45
The Cross Seizing in Net Making	46
On Making Jumping Ladders	47
Fixing the Tulip	48

CONTENTS

The Bulldog Grip or Clip 49
The Standing Turk's Head 50
To Finish off an end of Wire not Required to be Spliced 50
On Making Wire Grommets 52
A Note on Stranding Wire Rope 53
The Perfect Wire Grommet 57
Another Method of Grommet Making 59
The Ribbon or Lifting Band 61

Marline-spike Seamanship

Tools

The Rigging Knife — Start off with a sharp knife. A sharp knife is, without a doubt, the wire and rope worker's best servant. Proper rigging knives are manufactured, but any knife will do, providing it is large enough, and possesses a keen edge. A blunt knife only hacks its way through, and tends to loosen the whipping or seizing, when serving the ends after knotting. You can see the knot loosening, and before you have completed the job, the binding, unseen by you, may work loose, and in no small way the efficiency of the job is affected.

Rigging Tools — I use the lightest and strongest I can get, but you can not do light work with heavy tools, and vice versa. I seem to think that the fewer tools one possesses, the more versatile one becomes as a rigger. Let us start with marline spikes, they are made in various lengths and girths, with and without wooden handles, round and flat, of best cast steel hand forged. I always have my spikes modified, i.e. with the point flattened out and slightly turned upwards. I find that I can pick the strands up more handily. The common cork is ideal for protecting spike points.

Serving Mallets — Made in various sizes of brass and wood. Another handy tool to have is the home made serving board, made from a stave of a wooden bucket or a piece of hard wood containing three holes, through which is rove the serving of yarn. This tool is ideal for yarn servings of rope only, large and small. A lot of fancy tools might be had, such as heaving and serving mallet combined, small reels for serving wire, resembling miniature hawser reels, which may be attached to the serving mallet.

The Rigging Screw — Used properly there is nothing so efficient for seizing in thimbles. The type of rigging screw as used by the professional rigger is probably outside the demands of the amateur, but a reliable substitute can be made from a discarded or old bottle screw slip, and the necessary modification will readily be seen. There is however just one trick worth remembering. The rigging screw has a tendency to slip

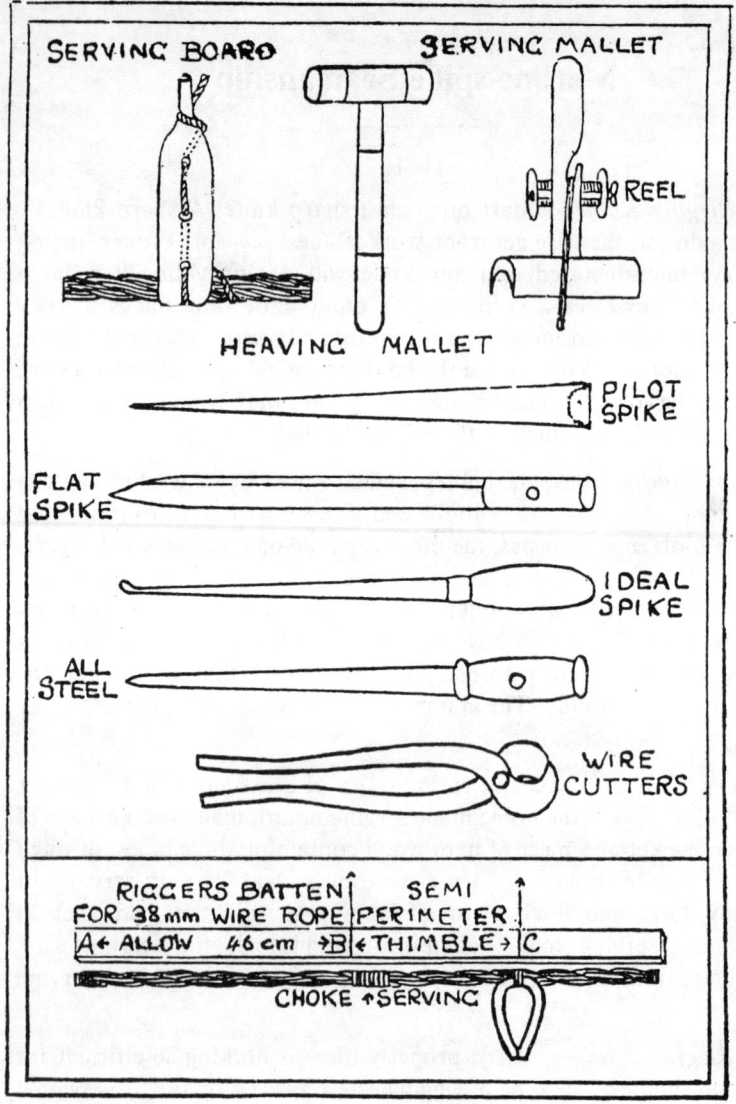

Fig. 1 — Rigging Tools.

downwards, or away from the toe of the thimble, during the act of screwing up. This snag can be overcome by securing a stout yarn to the arm of the screw, then hitching it to the crown of the thimble, and then on to the opposite arm of the screw.

Wire Cutters — There are many kinds of wire cutters, hand and hydraulic, etc., but when cutting the small wires, the hammer and cold chisel will suffice, and lest we forget apply a good whipping on either side of the point of cutting. A pair of end on cutters are needed for snipping off the yarns of the extra hard wires, and a pair of ordinary cutting pliers are very handy, a vice of course being indispensable, and if you are unable to secure the hollow jawed vice, use a pair of lead slabs in the ordinary type, for lead grips well, and being a soft metal does not injure the wire.

The Rigger's Batten — The batten is simply a thin lath of wood marked off at distances for the choke serving and crown of thimble. For various sizes of wires (Fig. 1), shows a batten for 38 millimetre wire. It enables the splicer to measure exactly the correct amount of wire for splicing, and the position for crown of thimble, and choke serving, thus affecting a considerable economy in wire.

Heaving Mallets — The best I know of consists of a brass tubular head and handle both of which are filled with a hard wood. It is a real labour saving tool, especially when passing the seizings, or when heaving taut the clove hitch. You can get just that much extra power on the heave, which makes so much difference to the tautness and efficiency of the seizing or hitch.

How to Use the Marline-spike

There is more skill needed in using the spike than is generally imagined. Once the using of the spike has been mastered, you will be well on the way to making an efficient splice. I have seen many points of spikes snapped off, and punctured fingers, caused by the bad handling of the spike, lay the point of the spike in the lay of the wire, press hard with the thumb of the left-hand on the point of the spike, simultaneously levering the spike towards the body with the right hand, taking good care not to fracture the main heart. The press and twist together can easily be accomplished with practice. Then force the spike smoothly through the wire, using your own judgement as to how far to enter it (Fig. 2).

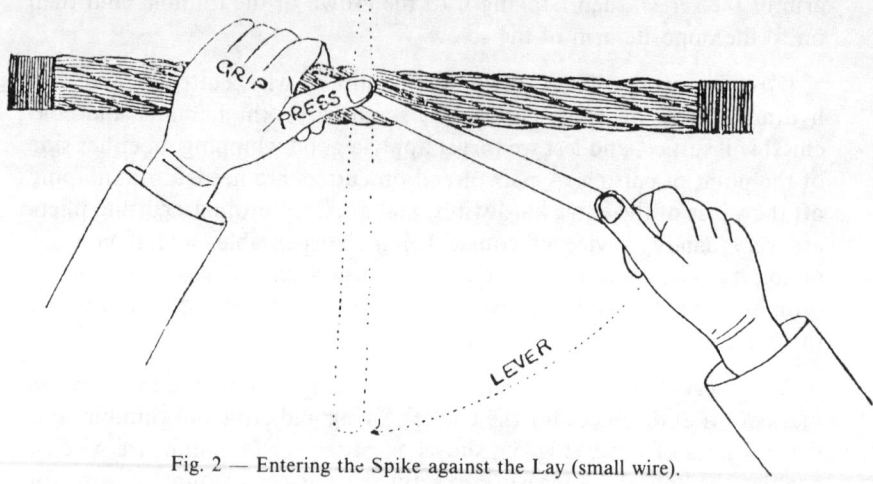

Fig. 2 — Entering the Spike against the Lay (small wire).

A few Words on Knotting

Always reef knot the ends. When using an extra large strand, say when seizing in the larger wires to the thimble, some difficulty may be met with

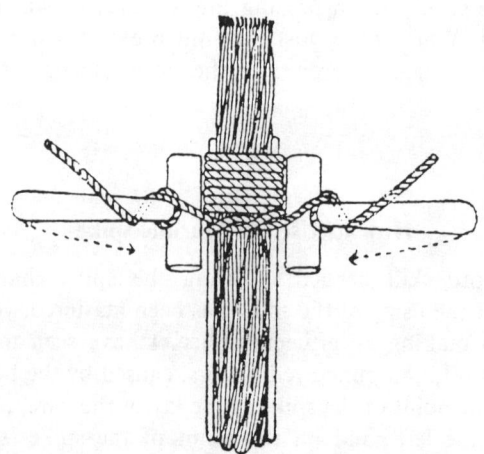

Fig. 3 — Thimble Eye, side view.
How to knot the larger bindings efficiently by heaving mallets.

MARLINE-SPIKE SEAMANSHIP

in getting the knot to finish sufficiently taut. I overcome this by applying two heaving mallets, one to each end of the strand, using the heaving turn, and haul taut each half of the knot separately (*See* Fig 3).

How to Serve

Always serve against the lay. For right handed laid wire or rope you can't go wrong if you have the wire under your left arm, and you are looking towards the eye and passing the serving away to the left from just below the last tuck and work towards the eye. The serving should be put on as taut as possible smoothly and without jerking the board. The best

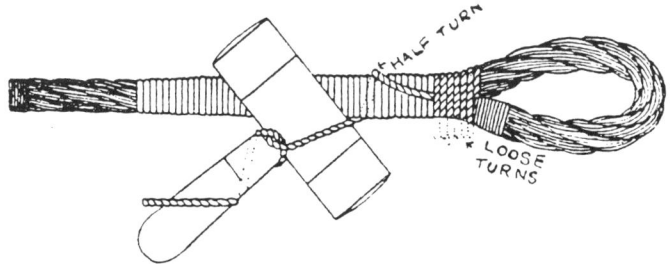

Fig. 4 — The finish of the yarn serving and turn for the final heave.
Note: The loose turns are hove taut singly by the mallet.

method I know of finishing off a serving of yarn is as for common whipping, i.e. work back loosely five turns, then reeve the working end of the serving under the turns. Now with the heaving mallet, heave taut the five turns separately, hauling the slack through, but do not apply the mallet at the point of issue, if you do, you are bound to lose the last turn

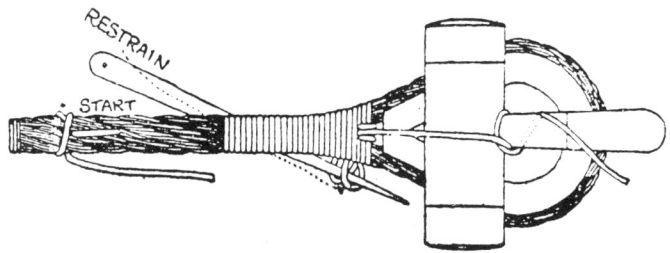

Fig. 5 — The start and finish of the wire serving,
via the cloth hitch, mallet and spike.

of the serving. Here is a little trick, take a half turn round the splice before applying the mallet for the final heave (*See* Fig. 4) cutting off 152 millimetres from the point of issue (Fig. 4). If the yarn is of three lay, make a crown not close down, followed by a wall knot below the crown and cut off. For two lay yarn reeve knot close down. The knotting ensure the ends not drawing.

Un-annealed wire of various sizes is used for the passing of wire servings. Commence by lifting one strand of wire just below the last tucks and enter the standing end of serving wire with the lay. Serve as already described and if the thimble is of the squared toe variety, finish off by passing a clove hitch completely round the last two turns of the serving, working the hitch taut in stages, i.e. half the hitch at a time, (*See* Fig. 5) but if the toe happens to be of the pointed variety I am afraid you will have to finish off by lifting one strand of the six wires which lay around the thimble, close to the toe, and taking two full turns round it and cutting off. I do not like this method of finishing as it spoils the lay of the wire at that point (Fig. 6). It will be found that the hitch cannot

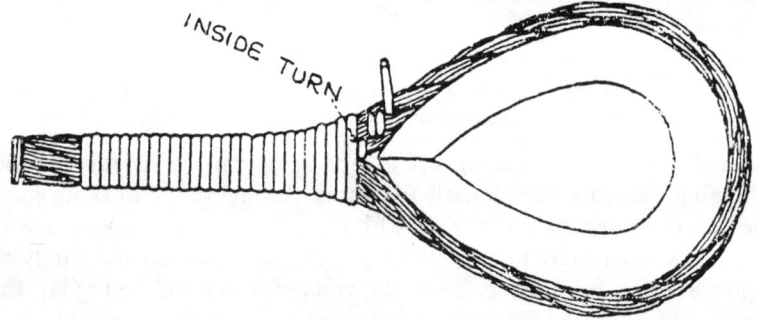

Fig. 6 — Another method of finishing the wire serving.

be hauled taut efficiently by hand. (Fig. 2) shows the hitch being hauled taut by way of the spike and heaving mallet as follows:— Enter the spike in the correct lower loop or bight, then apply the mallet and heave a steady taut, binding the serving wire on the spike, which draws taut the top loops. Now retire slowly the spike in a rotating movement to the very point, at the same time keeping on the pressure with the mallet to haul taut the loops, cutting off close to finish.

I have often thought that a serving should be commenced from the eye and finished beyond the last tucks.

MARLINE-SPIKE SEAMANSHIP

In serving in this fashion you surely do squeeze the tucks into the splice as you serve along it, as against a tendency to lift the tucks when serving towards the eye. This is very noticeable in rope. I always serve my soft eyes in this manner. With the thimble eye it is somewhat difficult to keep the first few turns taut and close and to the toe of the thimble but when serving over the short distance covering the ends of the strands in the spiral eye, you should have no trouble.

NOTE:—It has been said that a serving may cover a multitude of sins.

How to Worm

Worming is simply filling the scores of the wire, in between the lays, with yarn, to give it a more rounded appearance. The old rhyme says "worm with the lay", but you can't go wrong here, for whichever way you hold or look at the wire, you must go with the lay. The yarn in use should be well greased to prevent moisture from entering the heart of the wire.

How to Parcel

Always parcel with the lay, and the idea of parcelling is to preserve the splice. Use a strip of old flax canvas, bagging or insulating tape. I used a strip 25 millimetres wide for 25 millimetre wire and so on, passing it as

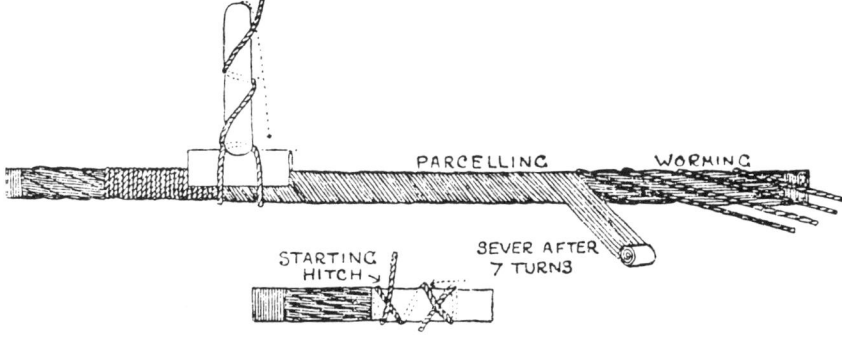

Fig. 7 — Worming, Parcelling and Serving.

taut as possible, in the same manner as a soldier puts on the now old fashioned putties i.e., each turn overlapping the other. Before parcelling treat the splice with a coating of stockholm tar or grease, then rub the parcelling, working the preservative well into it.

MARLINE-SPIKE SEAMANSHIP

On Whippings

The efficiency of the whipping on the end of the strand playing an important part in the splicing, a short passage, on the subject, will not be out of place. There are four, the Common, American, West Country and the Sailmakers. We only use the Common Whipping on the end of a strand. A whipping should be made with the smallest and strongest twine available, taking into consideration the circumference of the strand. It is obvious that a large or thick yarn is absolutely useless. It makes the end of the strand much larger in circumference, and makes it difficult for the splicer to enter the strand (I have heard lots of cussing over this). Waxed sail twine is as good as anything I know for the whipping of the larger strands. Haul each separate turn as taut as possible and terminate the whipping as near to the end of the strand as possible. The length of the whipping should be approximately the circumference of the wire of which the strand is part and parcel. (*See* Fig. 8).

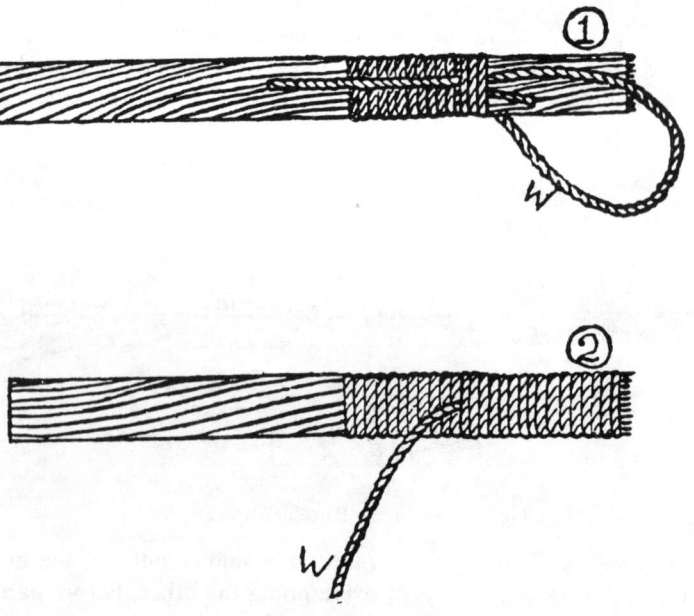

Fig. 8 — Whipping the Strand.

MARLINE-SPIKE SEAMANSHIP

How to Make up a Hank of Spurnyarn for Serving — Commence by taking half a turn round the shaft of a hammer, or the like placed in a vice, and then round the index finger, then continue taking the turns round the finger and hammer shaft diagonally, working away from you to the left on the shaft, and towards you and to the right on your finger. When enough turns have been taken, slide hank off the shaft and finish off with clove hitch round the centre of hank, with the end which now becomes the standing end. It will be found that the working end comes away freely when pulled enabling the server to work single handed, one hand on the mallet and the other hand to toss the hank over the wire as he continues to pass the serving (*See* Fig. 9).

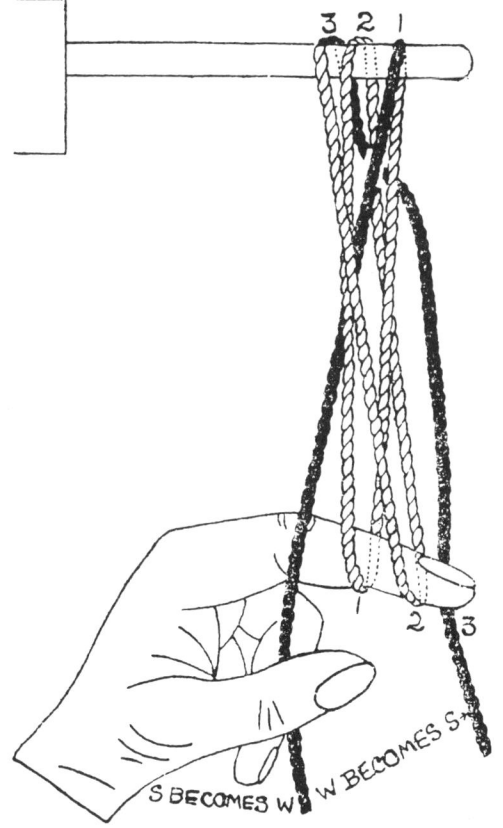

Fig. 9 — Making up a Serving of Yarn.

How to Join a Broken Serving — It sometimes happens that whilst serving you apply rather too much pressure on the serving mallet which results in the serving parting. I use the following method for making the perfect rejoining of the two ends. Work back about seven turns of the serving remaining on the splice, then lay the other end on the splice and serve over it, about seven turns and with that part take a turn round on the part laid on the wire, then carry on serving over that end, cutting off when you are satisfied that you have taken enough turns to bind it securely and compete as for serving (*See* Fig. 10).

MARLINE-SPIKE SEAMANSHIP

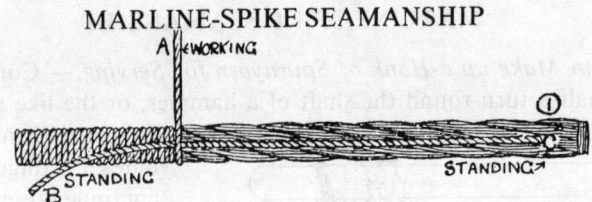

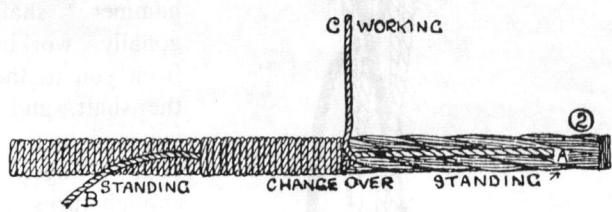

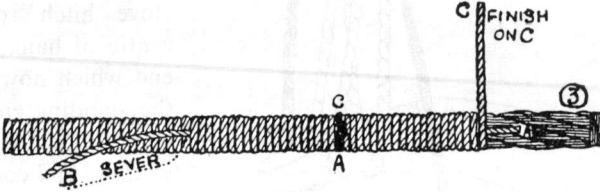

Fig. 10 — The perfect joining of a broken serving.

Fig. 11 — Making up the wire serving.

How to Make up Serving Wire — Commence by making a series of vertical loops with the wire. I usually make eight loops of about 152 millimetres in length, holding them close together in the left hand, then, with the right hand, expend the remainder of the wire round the loops, criss cross fashion. Now bend the tops of the loops down towards the hank. This prevents the criss cross wires slipping off whilst serving (*See* Fig. 11).

MARLINE-SPIKE SEAMANSHIP

Hints on Splicing in the under and over style

When splicing wires of the larger sizes, say 76 millimetres and above I enter and lift the strand with a small spike, driving it in with the hammer, then insert the larger spike, and withdraw the small. A good thing to remember is that wire, and especially large wire will have its own way, therefore it needs coaxing. Take care that you have no turns in the wire before splicing; I refer to the standing part. You can ensure safety in this respect by taking a turn out, e.g., opposite to the lay of the wire. When splicing the larger wires, hang them up waist high.

A good method is by swivel hooks with a lanyard in each end. By this method the wire can be turned over as you continue tucking. With small wires I sit down and lay them across my knees, and now that the days of bragging about corns are over I use a light pair of leather gloves, with thumb and fingers cut off. I enter the strand under and before the spike (free of turns) (Fig. 13), and if it looks as if the strand is going to kink, do not let it, withdraw it and have another shot, for once a kink is there it stays there. Now pull the strand hard on the spike then withdraw the spike and give the strand its final seating. The strand should be pulled through at right angles to the wire.

For big wire, seize the end of the strand with both hands, push it away from the body and jerk it in. I know some splicers who enter the strands below the spike, having previously rolled the spike away 25 millimetres along the wire, then rolling the spike back, take the strand with it to its final seating. Different splicers, different methods, but use the method which suits you best.

Just a few words on the locking tuck, I know quite a number of riggers, who insist on a locking tuck in all their splices. I describe a locking tuck thus:—Two strands of the working part, passing under one strand of the standing part, in opposite directions, both strands to be entered before removing the spike (Fig. 13).

How to Splice extra Special Wire in the under and over style

Extra special Wire is of an all wire construction, i.e., the strand and main hearts are of the same quality as the wire, and not of jute as in F.S.W.R. In splicing no hearts are removed. Simply tuck the main heart through the centre or under three strands against the lay. (You can identify the main heart inasmuch as that it has no natural lay), it is straight, then where it comes out lay it up in that continuing score, and

MARLINE-SPIKE SEAMANSHIP

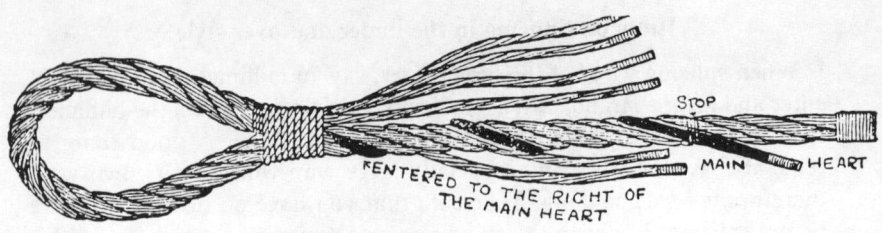

Fig. 12 — Splicing the 100% wire rope.

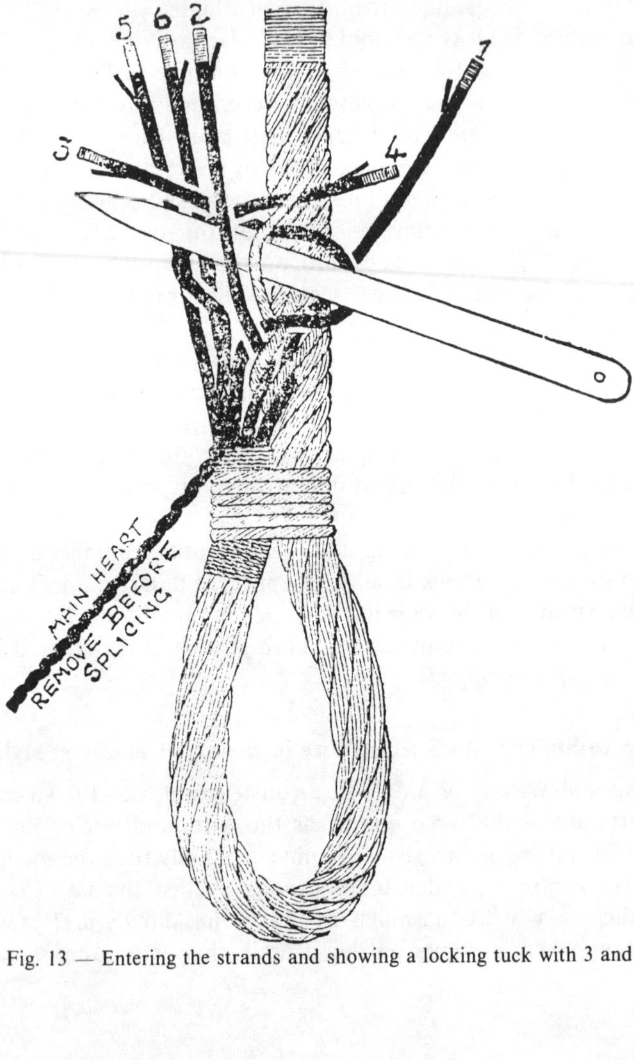

Fig. 13 — Entering the strands and showing a locking tuck with 3 and 4.

MARLINE-SPIKE SEAMANSHIP 13

stop the end down taut to the wire. Then splice as for ordinary wire burying the main heart as you splice, tucking over it as it lays in the score, remembering to separate the strands for the second row of tucks. (Fig. 12).

Method of Preparing Wire for Splicing in the under and over style

(The following instructions are for right layed wire rope having seven jute hearts).

THE SOFT EYE

The amount of wire required for splicing, naturally depends on how many tucks you propose to make in your splice, but where safety of life is concerned, not less than three whole tucks should be employed, on the other hand, e.g. when fitting a ship's guard rail, two and a half tucks would be sufficient. The professional rigger would perhaps be allowed only 229 millimetres for each millimetre size circumference of wire for splicing. But when employing the three whole, two thirds, and a third method, 30·5 centimetres should be used.

Example — For 51 millimetre wire allow 61 centimetres for splicing. At 61 centimetres from the end put on a good choke serving of 51 millimetres. Then bend the wire over to the standing part to size of eye required, and seize both parts together with a stout seizing. Then about 30·5 centimetres from the end of the working end, put on a light seizing. The object of the light seizing is to prevent the whole strands from unlaying whilst being whipped (Fig. 14). Now cut off the whipping on end, and whip the strands with an ordinary common whipping, allowing about 51 millimetres of the heart strand to protrude, then cut off the light seizing, and the next operation is to remove the main heart, by hauling it close back to the choke serving with the heaving mallet, and cutting off as close as possible.

Now arrange the strands three above, and three below the standing part. I might mention here that I find the weft pulled from old flax canvas ideal for whipping the strands of very small wire, and the yarns from white hemp the best for breaking in thimbles. Alas! we never see white hemp these days.

We now come to the operation of tucking the strands in the following sequence. Commence with the right hand strand of the upper three,

14   MARLINE-SPIKE SEAMANSHIP

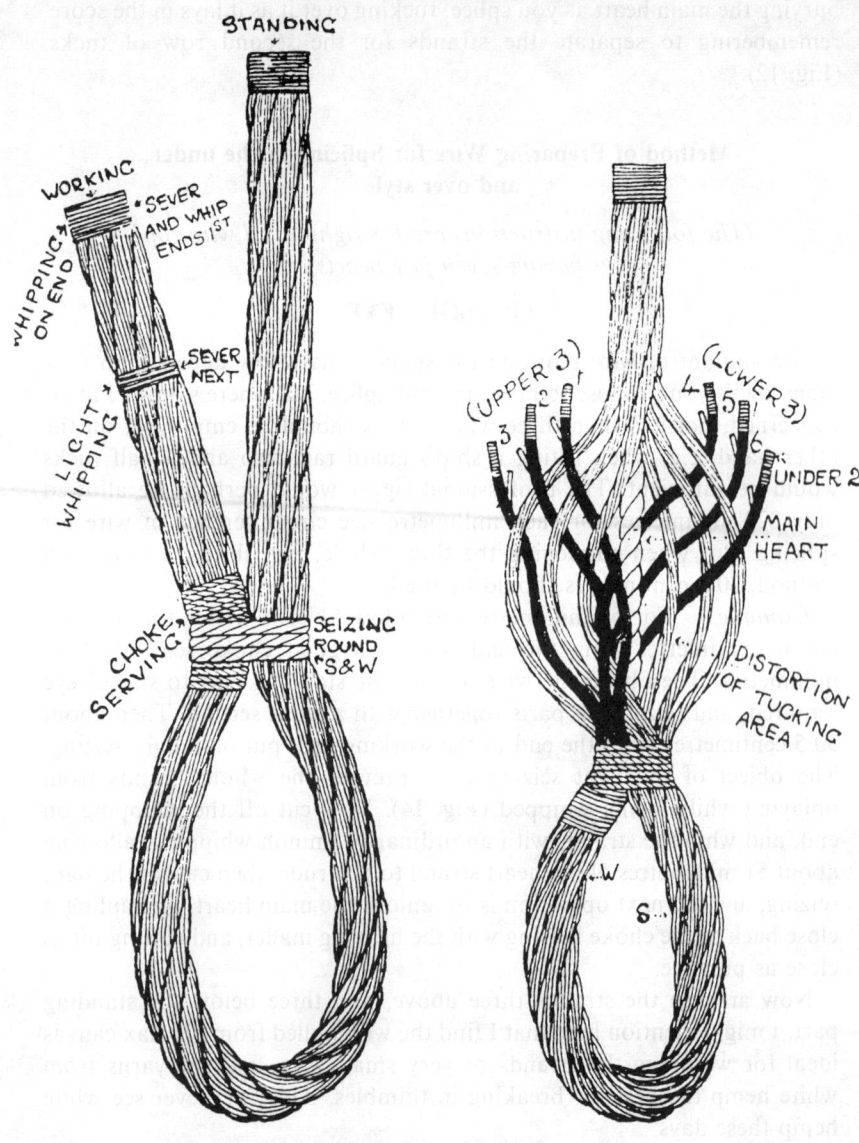

Fig. 14 — Preparing the soft eye.   Fig. 15 — The soft eye (first tucks).

which is number 1, but your number 1 must be selected correctly (*See* Fig. 15). It should lie immediately on top of the strand you intend to lift. A common fault is to take number 1 strand too far over to the right, causing a lump which cannot be knocked down. I maintain that, if the working and the standing parts of the soft eye are secured closely together, number 1 strand will show you where it wants to go.

Lift the strand and enter number 1 against the lay, that means from right to left.

We tuck number 2 next, also against the lay, and the same with strands 3 and 4. We miss number 5 strand, and tuck number 6 under number 2, then return to number 5, and then tuck it between the two strands under which number 6 passed. Thus we have all strands tucked against the lay, all strands over one and under one, with the exception of number 6 which passes under two strands. Number 6 is passed under two strands to eliminate the lump, which arises at this stage if all six strands are tucked over one and under one.

We have now one complete row of tucks. They should be well hauled back, and a sizing of light yarn put on, just above the tucks. This prevented the first tucks from lifting whilst you are performing the second tucks, and a seizing should be applied after each row of tucks is completed. We now come to the second row of tucks, and these we tuck, all six strands over one and under one against the lay having previously removed the hearts from all six strands. Then lightly hammer the tucks down. They have no friends, but do not beat them unmercifully, as you are liable to fracture the wires.

The over one and under two method can be used for the second row of tucks, and so on. It certainly gives a nice rounded appearance to the splice. I sometimes turn it in when splicing the smaller wires but you may find it rather tricky when using it in the larger wires owing to the short nip which arises after about the third strand has been entered.

Number three row of tucks, repeat as for number two. We now come to the attenuation, or the tapering of the splice which is as follows:—Separate the yarns of the strands one by one, take away one third of each strand, and tuck the remaining two thirds as for the previous tucks. Now separate the two thirds, and tuck one third. In the tapering tucks, care should be taken not to enter the spike too far. Remember you have only two thirds, and one third of the strands to tuck, and on completion they are liable to be loose. We now break off the yarns, by bending them down and twisting them down and twisting them to the right, which if performed correctly, tends to form a hook, and this looks much neater than cutting or chopping them off.

We have thus completed the splice as follows:—Once whole, twice without the heart, two thirds of the strand once, and one third once. — In all five rows of tucks, a simple but efficient splice.

The Hawser or Towing Eye

This eye is simply a soft eye, into which is introduced and seized a thimble, and is fitted in the ends of towing pendants. For should the thimble become distorted in a tow owing to rough usage, the seizing can be removed and a new thimble inserted. It also simplifies the splicing of the big wires. To find the size of eye required measure with a length of twine one and a half times the perimeter of the thimble plus the circumference of wire. Bring both ends together and you have the size of the eye.

Commence by making the soft eye and apply serving on its lower parts to prevent chafe at the toe of thimble. Then seize in the thimble, and you must apply the round seizing before you serve over the splice, for in this case if you serve over the splice first and then apply the seizing, you will find that the last few turns of the serving become slack, because they overlap the junction of the splice (*See* Fig. 16). It may be mentioned here that the splice of an extra large wire may appear lumpy or irregular. A more rounded appearance can be given by laying oakum in the hollows before parcelling and serving. (Or is this cheating?)

Attenuation — When using the under and over style, I favour the following attenuation:—Take away one third of the strand, but make it the innermost, or the most underneath third of the strand, these yarns are then twisted off, and the remaining two thirds are entered over the slightly protruding ends, to cover or bury them. Then repeat the operation when halving the two thirds and entering a third, the nett result being a splice so smoothly finished, that you could, if you so wished, and without fear, run the tongue along it.

Another method of attenuation is what I call a hurry-up job, i.e., tuck only three strands over one and under two.

MARLINE-SPIKE SEAMANSHIP 17

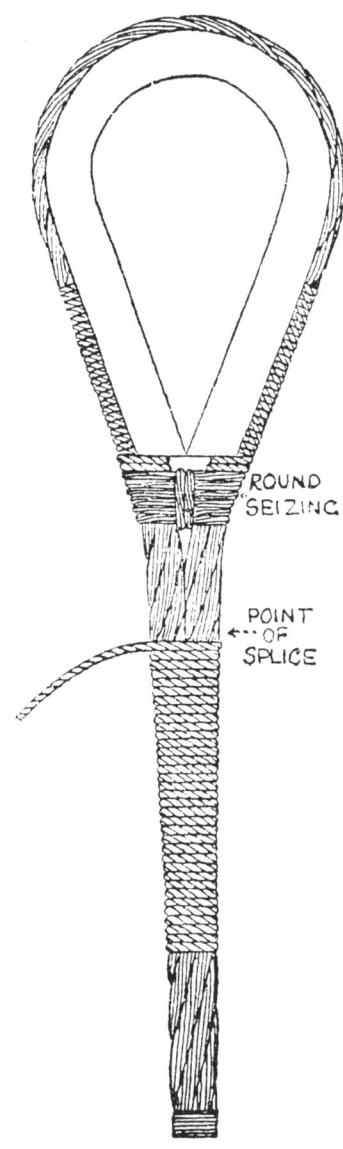

Fig. 16 — The Towing Eye.

The Spiral or Liverpool Styled Eye-Splice

A rather neat and compact splice, and the one which appears to be universally the favourite amongst riggers. It is quickly turned in, especially when fitting eyes to the larger wires. The reason is as follows:—

Once the spike is entered under a strand of the standing part, it is not removed from under that strand, until a full series of tucks have been entered, with the strand in use of the working part. The working strand is wrapped, in spiral fashion, round its counterpart in the standing part, whereas in the under and over style, the spike has to be removed and re-entered, before a further tuck can be made, the number of tucks made being not less than five and four alternately. The termination of the tucks odd and even gives the desired tapering of the splice. (It matters little whether you choose to complete number one strand odd, or even, if the first strand is finished odd, six must of course be finished even). A finer tapering can be arrived at by removing the hearts from the strands after the first tuck has been entered, and further, by snipping off a few yarns from each strand, after each additional tuck. Personally I think a better performance is made if the strand is left full or intact.

To Prepare for Splicing — The preparation no doubt takes a little longer than some other methods, and when splicing small wires, the extra time seems hardly warranted (but what is lost on the swings is regained on the roundabouts). For when splicing large wire you will turn in a splice, at least as fast, as when using under and over style. It is in fact a pleasant task. When using wires of less than 76 millimetres, allow 229 millimetres on end for each 25 millimetre circumference of wire for splicing, but when using wires of 76 millimetres and above, allow 30·5 centimetres end for each 25 millimetres. (Five tucks can definitely be made when using this formula). The removed main heart is kept handy of the wire is of 76 millimetres or above, for at a later stage during the splicing it plays an important part. The preliminary preparation of the splice, such as seizing in of the thimble, whipping the strands, etc., being completed, the eye is then hung up at a height convenient to the splicer, and in such a manner that the eye cannot move or turn during the act of splicing. Next a weight such as a common iron sinker or shackle is secured to the standing part, clear of the ground. It should be heavy enough to keep that part taut and in position, allowing the splicer to roll the spike up and down the wire, with the minimum amount of movement. The strands are then tied back and out of the way, in the correct order of

MARLINE-SPIKE SEAMANSHIP

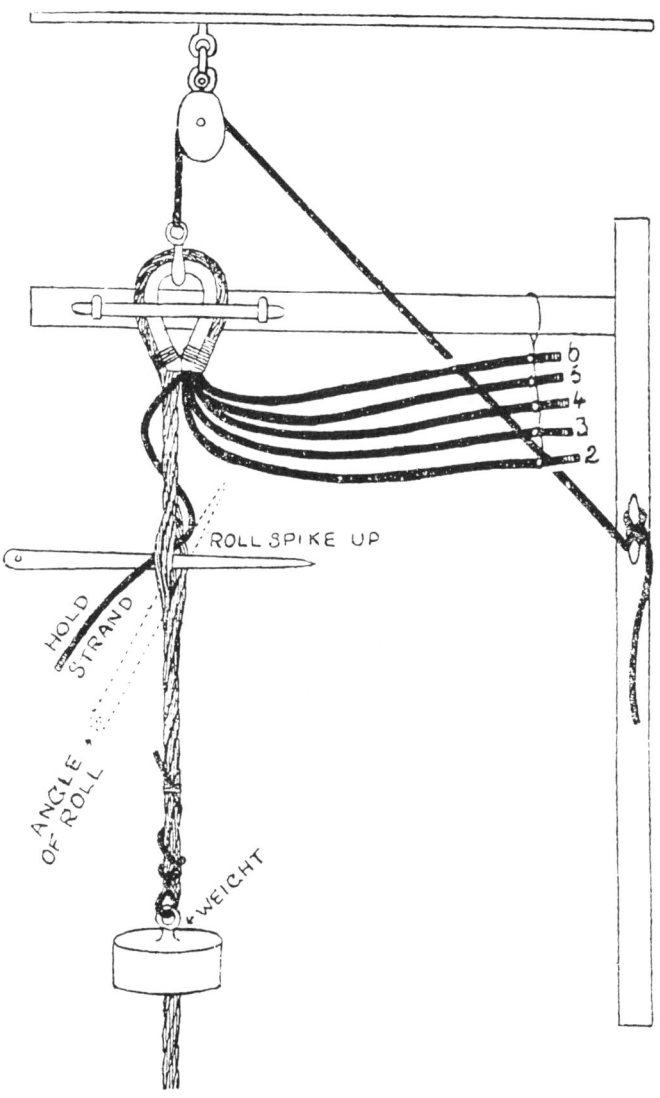

Fig. 17 — The Liverpool or Spiral Eye Splice.
Showing the preparation and how to enter number one strand.

tucking, with a light yarn, and in such a manner that they can be taken down singly without disturbing the remainder, but leave number one strand free. (Fig. 17).

How to Splice in the Spiral Fashion — Having selected the correct number 1 strand, tuck it under three of the standing part. It is very important that the correct three are lifted, for the correct entry of the first strand is the secret of all good splicing. The best way I know of securing accuracy in this respect is as follows:—Sight your entry of number 1 strand into standing part, then with the finger trace that continuing score down the standing part about 30·5 centimetres. Then count three strands to its left. That is the position for entering the spike, and the correct three to lift. I think I have made the perfect entry with a number 1 strand, if, supposing I removed the choke serving on the completion of the splice, the lay of the wire in that area remained virgin. In big wire do not attempt to lift the three strands together, try two first, then with the aid of another smaller spike, pick up the third. The spike should enter the wire, three to the right of the main heart, or in other words, with the main heart plus three strands lying in the left of the spike.

The Spiral Eye

(*See* Fig. 17.)

A round turn must be taken with the strand on the standing part, before it is entered, this is necessary in order to get the desired length of roll. (For without the round turn, there could be no roll when taking the strand home to its final seating in the splice). This means that the distance between the spike on entry, and the toe of the thimble is approximately equal in distance to one complete revolution of the spike, and in that ratio, one round turn will suffice. If on the other hand, however, after entering the spike, it needs say two revolutions to arrive home, and you have only one round turn with the strand, you will find that you are unable to complete the full length of roll, and force the strand home. So the point to remember is this:—That for the number of revolutions needed with the spike, when rolling home, a similar number of round turns with the strand are needed on the standing part (splicing area).

NOTE:—When rolling the spike, it should lie in the wire, with the point lying parallel with the lay and not at right angles to the wire. It will be readily seen that during an upward roll, the amount of spike entered decreases, and that during a downward roll increases or, in other words, the wire creeps off and on the spike. This is of course due to the lay of the wire. This is what I meant to infer, when referring to the use of too much spike. This can be remedied, by watching the spike during the roll, and adjusting it accordingly.

NOTE:—The strand is entered with the lay.

After the spike has been entered roll it close to the toe of the thimble, and take a sighter with number 1 strand for line of entry, and if correct roll down the spike, back to the starting position. Then take number 1 away to the left, behind the standing part, and enter it with the lay, and above the point of the spike. The end of the strand passing in front, and not in the rear of the spike. Just before entering the strands, slightly bend them to their natural lay in the splice, (try to visualise this). The strand having been fully entered, now hold on firmly to its end, and with the other hand, roll the spike up taking the bight of the strand to its final seating. Let the business end, or the point of spike do the work of forcing the strand home (it develops good spike seamanship). Some have a habit of tapping the point of the spike up with the hammer to force the strand home, which simply ruins the spike. The first tuck is then locked as follows:—Hold the strand in position and roll down the spike. The end of number 1 is then rove through the thimble, secured by a stop and for the moment forgotten. Next pick up the second and third strands of that three, which means that your spike enters under the next strand to the right of number 1, but, re-appears out through the initial entry of one strand and actually touches one strand. Then roll down the spike, enter number 2 and rolling it as close as possible to number 1, lock it.

Now comes a little manipulation, for one of these two strands has to be dropped, to allow the full series of tucks to be made with number 2 strand, and the spike artist does it so:—He just works the spike back to its very tip, drops one strand, and with just a flick of the spike, picks up the innermost strand, and proceeds to enter the full series of tucks, by an evolution of rolling and locking. Number 3 is entered next, under the strand just recently dropped, and rolled as close as possible to number 2, then add the full tucks. Next enter and complete the tucks with number 4, after which when splicing wire of 76 millimetres and above, the strands of

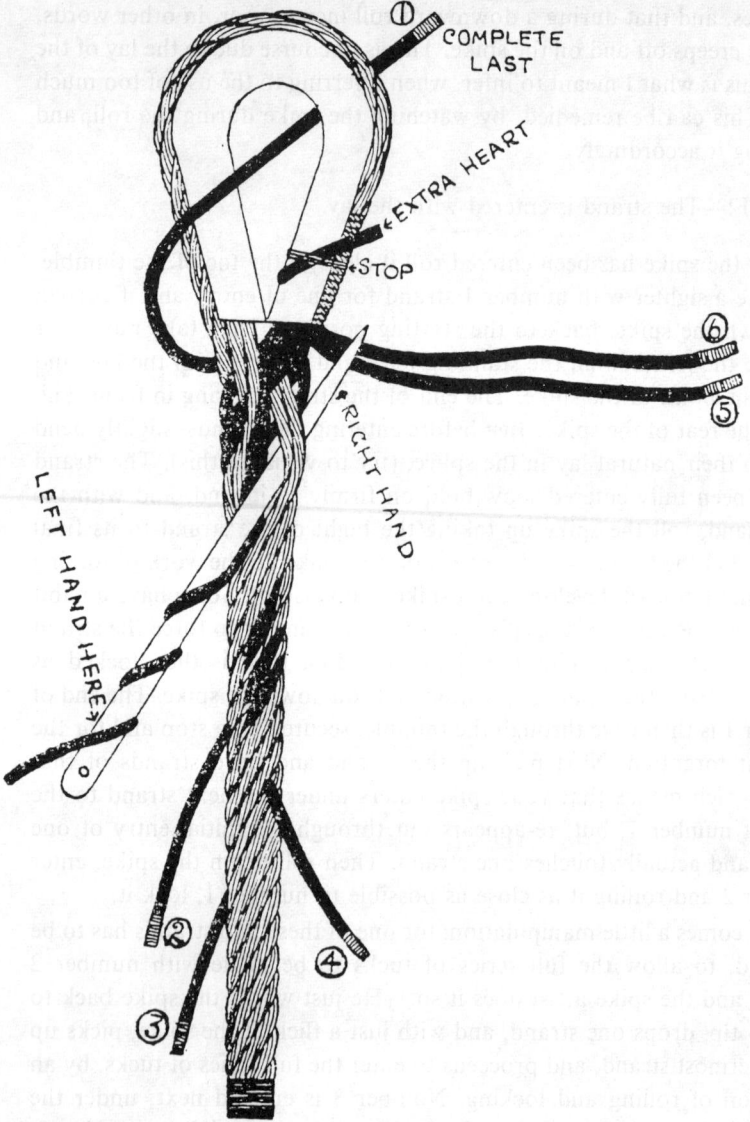

Fig. 18 — Entering the additional heart after the fourth tuck.
NOTE: Roll the spike down.

the standing part have a tendency to sink or drop, making them rather difficult to lift. This is sometimes due to the entry or use of too much spike when rolling the strands, therefore the least amount of spike used the better. To kill this complaint two thirds of the removed main heart is rolled into the unfinished splice. With the heart of three strands, it means of course the removal of one. It is rolled into the splice thus:—Lift the remaining two strands, 5 and 6 of the standing part, and enter one end of the heart through and secure it to the thimble. Now with the other end take three round turns, left handed, on the spike (*See* Fig. 18) and rolling the spike down, drop in the extra heart, the full length of the splice, and cut the lower end off close. Then complete the full tucks with five and six strands, and then turn to number 1 strand, and complete and lock the splice, by entering the additional tucks. Then trim off the top end of the extra heart, and lightly top the splice into shape. This splice being very compact, it is only necessary to serve it over in the area covering the protruding ends. When fitting a soft eye in this style, apply a bulldog grip on both parts, just above the point of splicing to keep the eye in position.

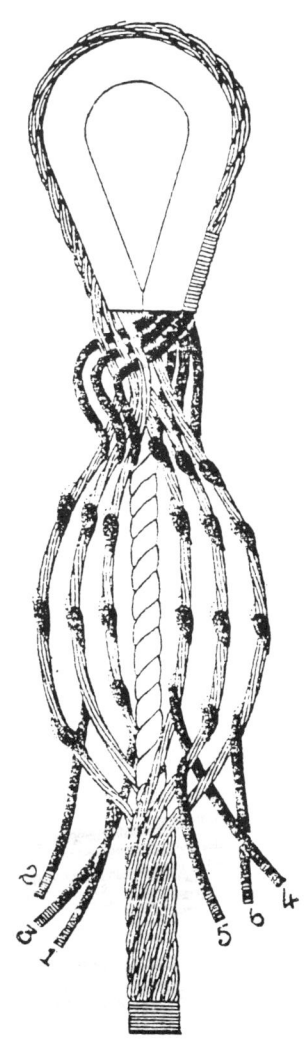

Fig. 19 — The Spiral Eye.
Showing how the tucks lie.

NOTE:—This type of eye is generally fitted to wire ropes that are not liable to spin. When splicing the larger wires, it becomes a job for two — one to work the spike, and another to enter and tend the strands.

MARLINE-SPIKE SEAMANSHIP

The Liverpool or Spiral Short Splice

Prepare the wire as far as the short splice in the under and over style, but instead of marling one half of the strands down to the standing part, apply a bulldog grip, extra taut, over the point of interlocking, next attach the weight to the standing part, and commencing with any one strand of the lower half, enter the complete tucks, five and four alternately, then working round the wire to the right, complete each strand one after the other, not forgetting to enter the extra heart, after the fourth strand has been tucked, if the wire is of 76 millimetres or above. Then reverse the wire and complete the splice, by entering the opposing six strands.

NOTE:—It is not advisable to remove the whippings at the junction, prior to splicing, as in the under and over style method. If you do, you will find that the rolling of the spike loosens the lay of the standing part considerably. Take care that the first strand you enter, does not enter under the strand of the standing part, directly in its line.

The Short Splice

IN THE UNDER AND OVER STYLE

(Right layed wire rope having jute hearts).

This splice is generally used in joining two wires together, not required to travel through a block, and in the construction of wires strops and

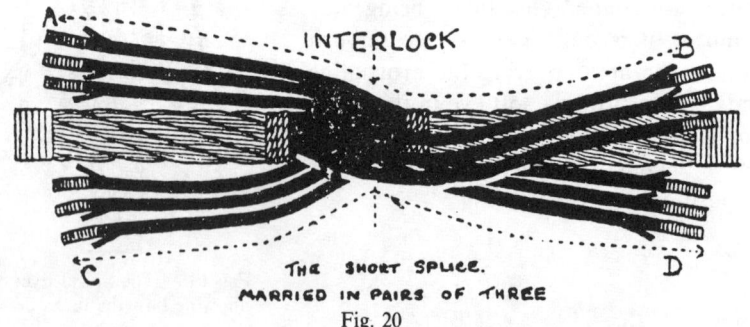

Fig. 20

spans. Prepare the two ends as for eye splice, allow the same amount for splicing and almost the same method of splicing, the only difference being that all strands are tucked under and over one. Commence by crutching, or interlocking the twelve strands together as in (Fig. 20). To

MARLINE-SPIKE SEAMANSHIP

simplify this, lock them together in pairs of three, first, two pairs of three in the opposite direction; now single them out to six up and six down, and force them together as close as possible, i.e., both whippings meeting each other. Then apply a strong seizing round the junction (Fig. 21). Now

Fig. 21 — The Short Splice, Singling up.

serve hard down to the wire one of the sets of six strands. It does not matter which six you select (*See* Fig. 22). Now remove the distance whipping to which the served down strands are unlaid, then tuck the unserved six strands the full series of tucks, and not until then do you turn to the opposite six strands.

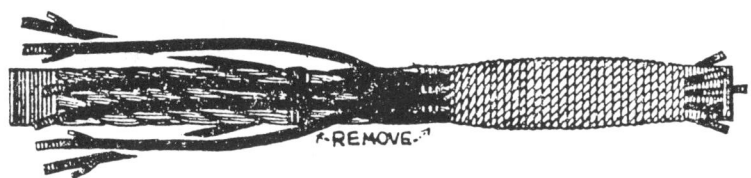

Fig. 22 — The Short Splice, Stage 3.

The tucks having been completed one way, remove the serving and the whipping from the remaining six strands, and tuck as previously described. The beauty of this method is, that when completed the long jaw caused by the distant whipping at the point of locking is eliminated, and it is exactly the same as the short splice in rope. But remember that the success of this splice depends on how efficiently you serve the six strands down to the wire. I repeat, serve them down as taut as possible. When making a strop or span, half a turn should be taken out of the wire before interlocking the strands, failure to do this results in the strop being twisted or with half a turn in it after splicing.

The Reduced Eye Type "A"

This eye is generally fitted in the ends of topping lifts, purchases of cranes, and the ends of all wires required to go to a power worked drum or winch, the distribution of the splicing giving flexibility, allowing the splice to bend the drum. Allow 91 centimetres on the end for each 25 millimetre circumference of wire and apply a whipping at that distance. Next remove the main heart and whip the strands. Now select three of the six strands and lay them up (laying up two first and then the third) and apply a whipping on these three, at the size of eye required. Then form the eye by bending round the three laid up and interlock them with the remaining three as for short splice, seizing, standing and working parts together just below the interlocking area (*See* Fig. 23). If you have chosen the correct strands to lay up, and unlay, there should be no half turn in the eye when completed, which is the common fault in this type of eye. Now comes the tricky part. Stop down to to the eye strands *B* and *C* which belongs to the set to be unlaid, leaving free stand *A*, then carefully cut off the distance for splicing whipping, and commence by unlaying *A* of the upper three, following it immediately in its wake by *A* of the lower three, leaving about 30·5 centimetres for splicing, and apply a good whipping at the crossing. (It will be noticed that the unlayed *A* strand will be considerably longer than 30·5 centimetres, and if it gets in the way snip some of it off). Then repeat the operation with strands *B* but

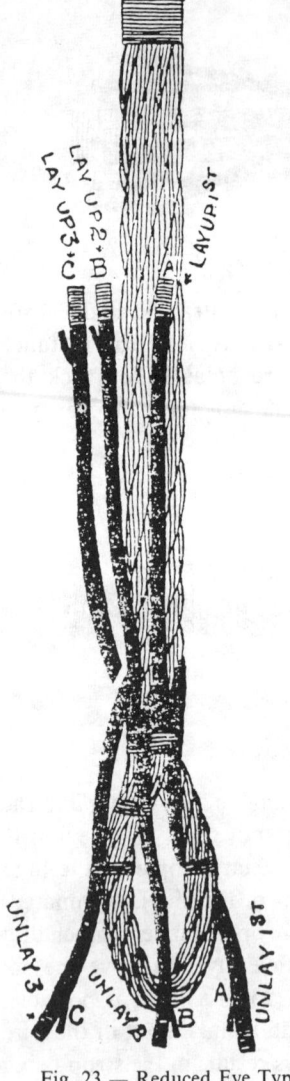

Fig. 23 — Reduced Eye Type. 'A' (Stage 1).

terminate the unlaying and laying, so as to finish midway between strands A and C when splicing is completed. The unlaying and laying up of strands C being completed now tuck all strands over one and

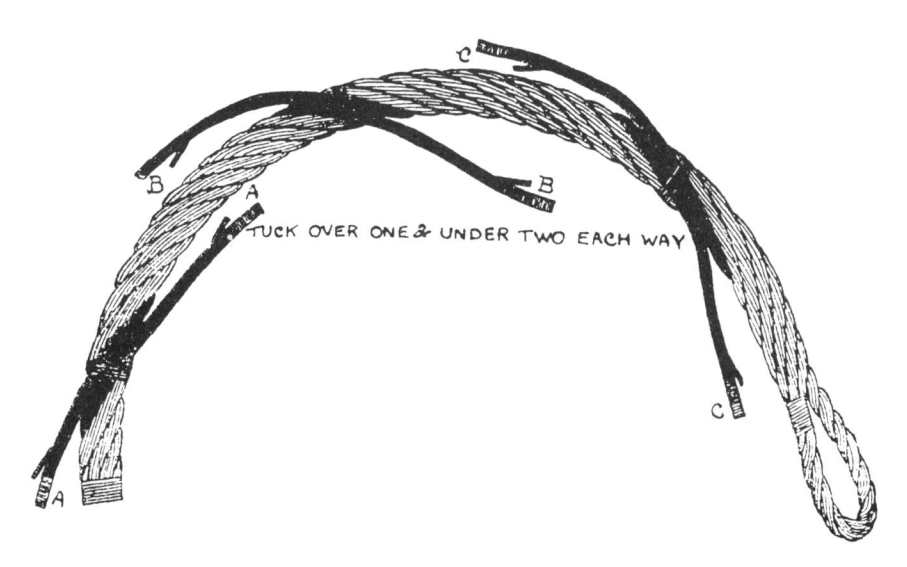

Fig. 24 — Reduced Eye Type "A".
Stage 2.

under two each way, then remove the hearts and repeat the tucking a second time each way. Complete by serving splice from C to A. It will be observed that more wire is required for splicing than for type B reduced eye (*See* Fig. 24).

The Reduced Eye Type "B"

This eye is perhaps not so neat in construction as type "A", but it is no doubt a stronger eye, with a simpler construction. Allow 61 centimetres on end for each 25 millimetres circumference of wire, and apply a whipping at that distance, then prepare the strands for splicing. Now take the innermost three strands and lay them up, then form the eye to the size required, and remove the hearts from the three strands, tuck the three strands as for eye splice in rope, three full tucks, Now tuck the upper three strands into the three laid up, this time with the hearts in until you meet the lower three strands. The tucking of the strands with, and without the hearts give a tapering effect. Complete as for type "A". To further strengthen the eye apply a flat seizing of unannealed wire. (*See* Fig. 25).

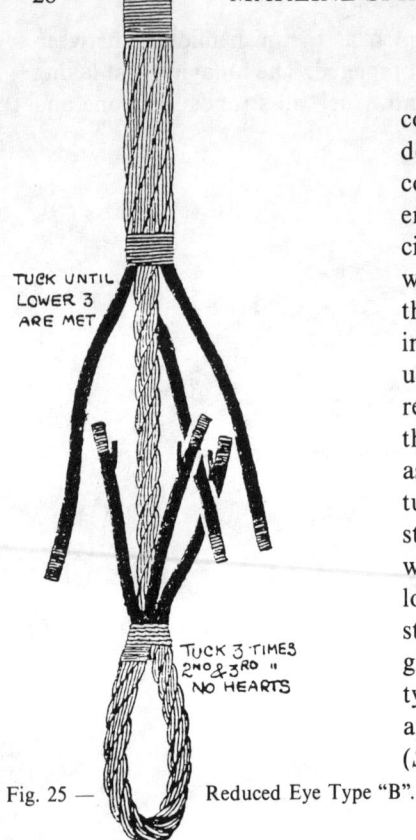

Fig. 25 — Reduced Eye Type "B".

The Flemish Eye in Wire

If you can produce a Flemish Eye of rope you should have no difficulty in producing one of wire. The construction is as easy, and differs little; and if you want to know why it is called the Flemish Eye, you will have to ask someone else. It is as difficult as answering a question such as: — Who was Matthew Walker of knot fame? But it really is a job worth knowing. It is a type of eye which I would not give a lot of work to do. It would be efficient enough, say, when fitted in a stay or shroud. Also it is the type of eye I would fit if, perchance, I found myself in the unfortunate position of being without a spike or the like. (I can hear the old shell-backs bursting forth with the old cry "Every finger a marline-spike, my lad"). It can be quickly turned in, and to all intents and purposes must be regarded as a makeshift eye.

It looks extra good when fitted to the flexible or truelay wires. (Truelayed, or as our American friends term it), performed, which seems to me the more logical term, means that the ordinary or regular layed strands, are kinked or performed to the required lay, by being worked under and over the metal sheaves of the transmuter head before reaching the final rope closing stage.

Do not attempt it with wires of iron or tough metals — you will not make it. The amount of the wire needed to form the eye, naturally depends on the size of the eye required, but as a reliable formula would not be less than 457 millimetres for each 25 millimetres circumference of wire.

How to Form — Measure off the amount of wire for splicing, and at that point apply a good whipping, and if the wire is of the common flex apply a light whipping of 25 millimetres or more from the whipping on end, but if the wire is truelayed, this light whipping is not needed for the strands will not unlay whilst being prepared. Then very carefully unlay three of the strands. Let us say, 1, 3 and 5, which in future will be referred to as "A" group, leaving, 2, 4 and 6, known as "B" group, undisturbed around the main heart. Next apply a whipping at the size of eye required on "B" group, and cut out that portion of main heart not required. Then form the first half of the eye, by bending "B" group to meet and interlock correctly with "A", hard up to

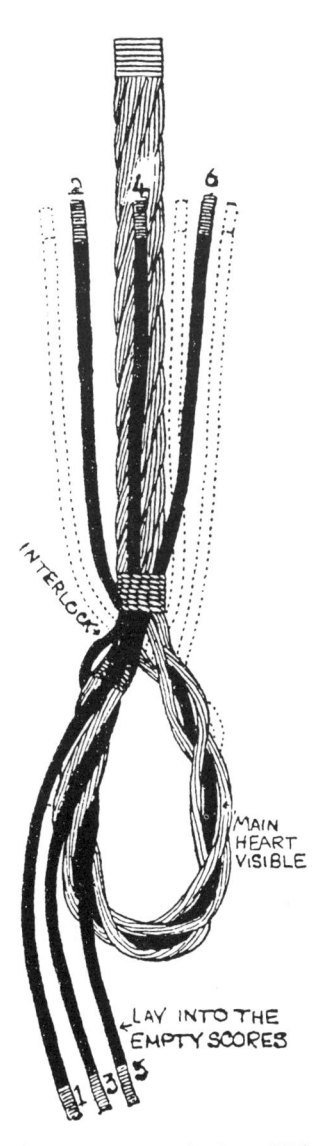

Fig. 26 — The Flemish Eye of Wire.

the distance whipping on the standing part. (Which is similar to the first stage in type "A" reduced eye. (Fig. 26). Then secure the ends of "B" group to the standing part. Next lay into the vacated spaces or scores of "B" group the strands of "A", after which all six strands should be pointing up and alongside the standing part. (*See* Fig. 26). Then single out the yarns, and marl and serve them down to the standing part. Complete by serving over the eye. A thimble may then be inserted and seized into the eye.

The Cut Splice in Wire

Personally, I can see no use for the cut-splice in wire. An old timer, though, assures me that it will, as regards service, outlive at least two long splices, and if the block will take it, may be used as a substitute.

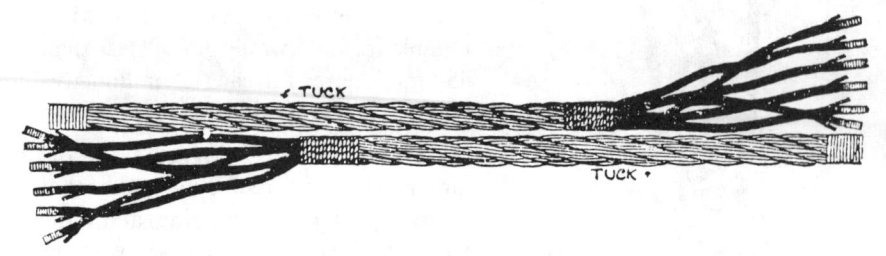

Fig. 27 — The Cut Splice.

To Form — Lay the two ends together and overlapping each other with enough end for splicing. Then simply tuck the strands of one into the other as for eyesplice. A flat seizing may then be applied in the centre of the loop. (*See* Fig. 27).

Splicing Wire to Rope

This splice has to be frequently made by yachtsmen, say for a wire with a rope tail. The snags here are that the wire strands are tougher and not so pliable as the rope strands, the wire lasts longer than the rope, and there are six strands of wire compared with three strands of rope. I use the following method and it has never failed me. The rope tail should be approximately the same circumference as three strands of the wire when laid up. For 38 millimetres F.S.W.R. apply a whipping at a distance of 76 centimetres from the whipping on end, then whip the six strands allowing the hearts to protrude, and carefully remove the same. Now lay up the

MARLINE-SPIKE SEAMANSHIP

innermost three strands allowing 45 centimetres for splicing. Then take the rope and allow about 30·5 centimetres for splicing, and crutching them together, short splice, giving four tucks each way, cut the ends off and roll the splice under foot. We have now three strands of wire remaining, and I treat them as follows. Single out the yarns and marl them down over the splice, snipping them off at various intervals to form a tapering of the splice. Complete by parcelling and serving over all as taut as possible, binding wire and rope together (do not forget the tallow). (*See* Fig. 28).

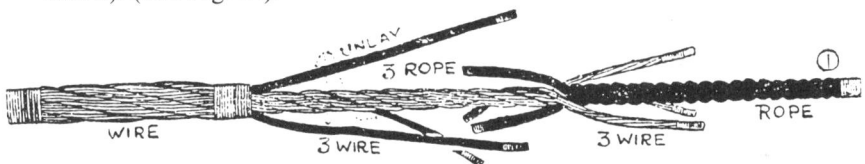

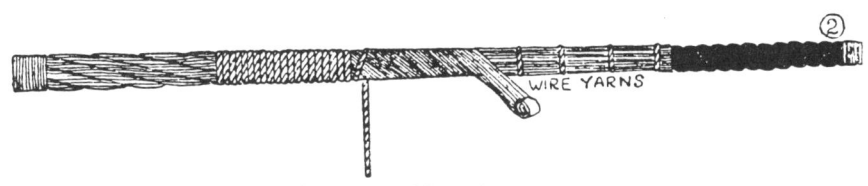

Fig. 28 — Splicing Wire to Rope.

The Long Splice

The king of all splices, and a difficult job to describe, the most discussed point of this splice is how much wire is required for splicing. I know of no hard and fast rule, but one thing is sure, that the longer the splice is the stronger it is. In any case the strain which the splice will have to bear is the most important factor.

I think that the best method of learning the long splice is by procuring two lengths of septin, which is a six-stranded French rope, but it is very difficult to find these days. The twelve strands of rope confront you with some small measure of the difficulty likely to be encountered with when dealing with twelve strands of wire. Do not try to lay and unlay the strands singly (most would-be long splicers think they can), if you do you will soon find out that you have boobed over the job, loosing the true and symmetrical lay of the wire. They must be worked in pairs.

I will now get on with the worst job of all and attempt to describe the long splicing of 51 millimetres circumference F.S.W.R. into an endless band, and to be on the safe side as to how much wire to use. I say ten

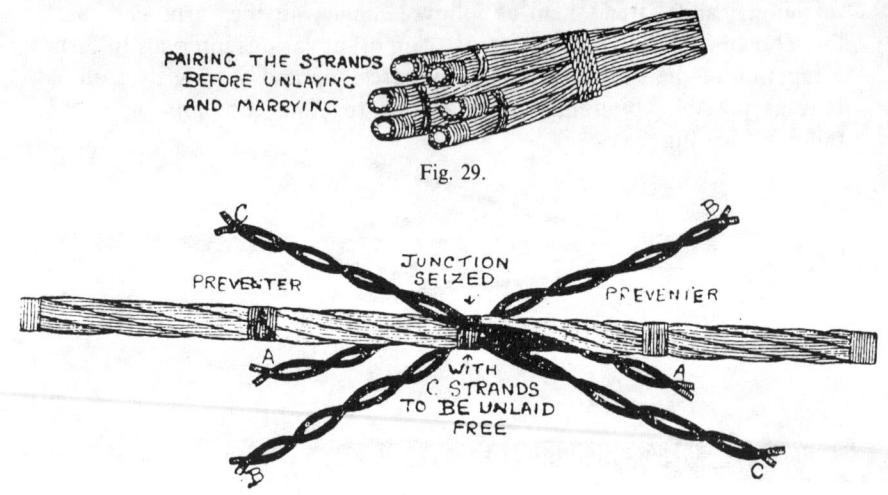

Fig. 29.

Fig. 30 — The Long Splice. Stage 1.

times the circumference in centimetres, which gives us 6 metres; so at 6 metres from each end supply a good whipping, and then about 152 millimetres from the whipping on the end put on a light whipping. Now remove the whipping on the end of each wire and whip all the strands. Then select the strands in pairs as they lay around the main heart; you can not go wrong here, any two will do, and seize them in pairs 25 millimetres or more from their ends. (*See* Fig. 29). Now remove the light seizing allowing the pairs of strands to unlay the 6 metres distance whippings. Next remove the main heart from each wire by hauling them as far back as possible with the heaving mallet cutting off as close to the distance whippings as possible.

Now marry the pairs of strands together as for short splice in cordage (*See* Fig. 29) and force the junction together (try to make the distance whippings meet each other), then supply a clove hitch of stout yarn around the junction and leave it as taut as possible with the mallet, but leave on the outside of the clove hitch, the pair of *C* strands which have to be unlaid (*See* Fig. 30). Next remove the distance whipping *C* half of the splice. Then proceed to unlay the free pair of strands *C*, following it in its

MARLINE-SPIKE SEAMANSHIP 33

wake, by its opposite number *C* working them well into the vacant lay to the distance required — say about 4·5 metres — and apply a good seizing of yarn at the junction. Now switch to the other side of the junction, remove the above hitch and repeat with *A* exactly the same as strands *C*, and the splice at this stage appears as in (Fig. 31), resembling the long splice in cordage.

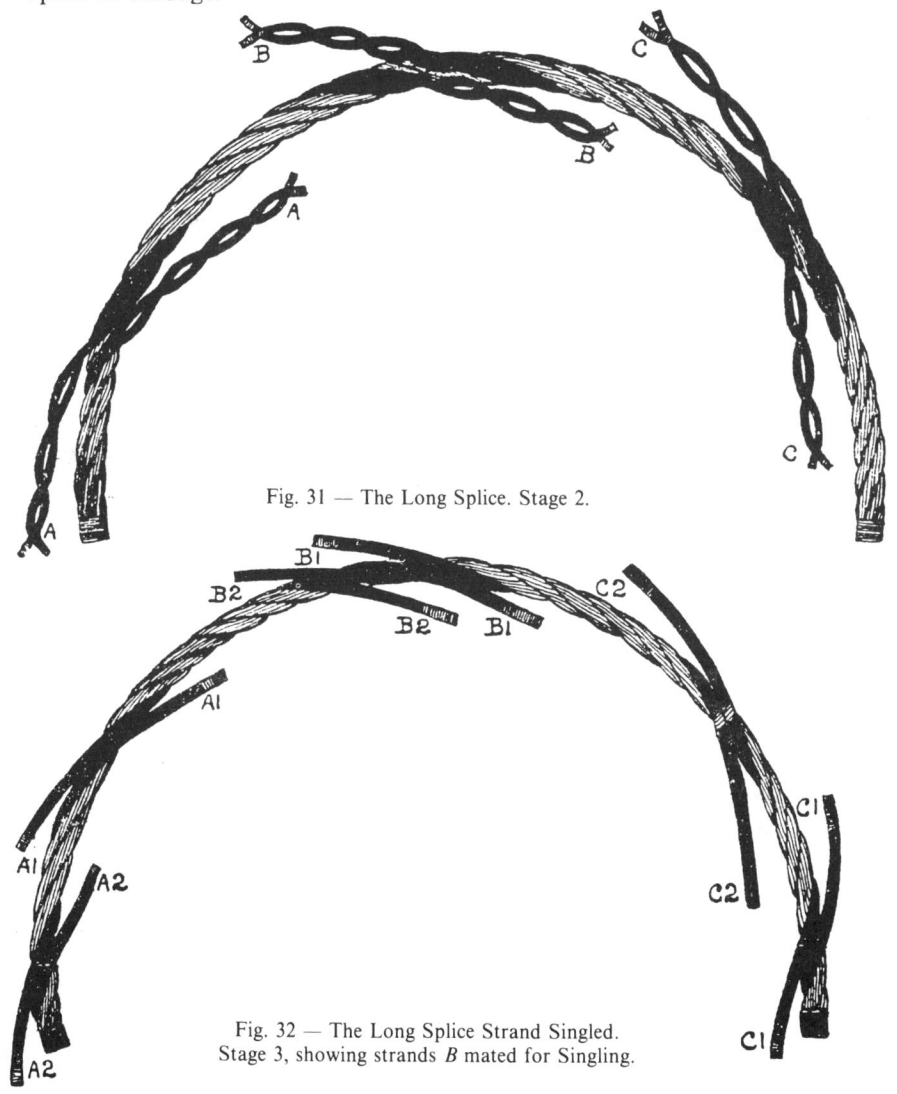

Fig. 31 — The Long Splice. Stage 2.

Fig. 32 — The Long Splice Strand Singled.
Stage 3, showing strands *B* mated for Singling.

We now come to the operation of singling the pairs and interlocking them (see also Fig. 32), taking care that they are properly mated, otherwise you will not get the true lay. Then commencing with strands *B* unlay and lay up, to approximately the correct distance, about 2·3 metres, but you must, I repeat must, when you decide to terminate your unlaying and laying up, apply a stout seizing at that point, otherwise it is sure that you are going to lose what you have up to the present gained. (Remember that you can always get the correct distance apart when you come to finish off). Then repeat the operation with strands *A* and *C*. The singling up being completed, we have twelve ends to get rid of (at this stage snip off excess wire of unlaid strands and whip ends), and this is done in exactly the same way as the finish of the perfect wire grommet. (*See* Fig. 33) taking care that, after entering the two ends through the centre of the wire, that you get the two parts lying evenly in the lay of the wire at that point.

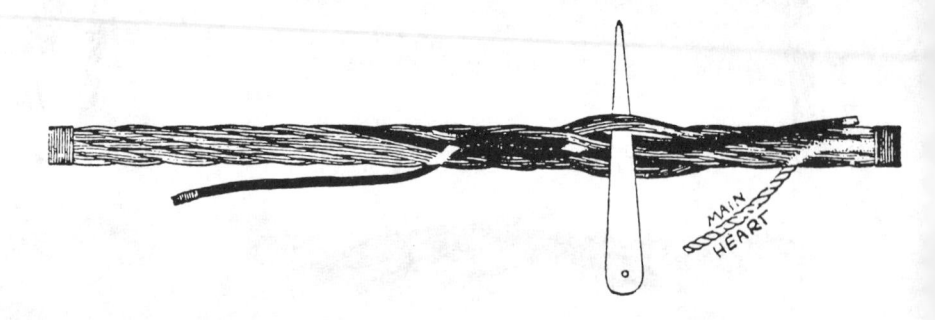

Fig. 33 — The Long Splice. Burying the Ends. Roll Spike Clockwise.

If you can do a good job of finishing the perfect grommet, you should have no difficulty in finishing the long splice. The only difference is that in the perfect grommet, you remove the pilot strand, which is its temporary heart, a little at a time and roll in the end of the grommet, whereas in the long splice, you remove the main heart with two spikes, sever and remove it some distance, by rolling it out with the spike, and then rolling in the end of the strand, but be uniform in the distance apart of finishing the strands, and the amount of end you roll in, in this case it should be not less than 152 millimetres, and see to it that the ends are efficiently whipped; a better method is to give the ends a coating of solder.

In small wire I roll the splice under foot, in big wire I lightly hammer the region of the entry of the strands into the heart of the wire with the mallet, and the strain involved on the wire does the rest.

It must be understood that in the long splicing of the larger wires, all tricks and kinds of dodges are exploited, seizings here; clamps there, and believe me when I say that all these manipulations come naturally to the keen and studious splicer, they stick out a mile. Remember the trouble you encountered when you first laid up number five part of the grommet, the difficulty you had in keeping it in the lay, well, you have the same trouble with the strands in the long splice.

It is a safe bet that you are not going to make an efficient long splice at your first attempt or the second for that matter, such skill is not so easily acquired, but remember that practice makes perfect, and if you get the opportunity watch the old timers at work in the rigging houses; they have had a load of knowledge, and they work so deftly.

Wire Seizings

Nothing looks smarter than a neat, efficient seizing, and they are a joy to perform. There are three. The round, racking and the flat. Never try to put a wire seizing on the bare wire both parts must be first served over with suitable yarn. Otherwise the seizing tends to slip as the turns are passed (the serving gives a good foundation). Let us start with the round seizing. Commence by twisting an eye in one end of the wire, and reeve the working end through it, and round both parts of wire to be seized. Then pass the required number of turns loosely; let us say eleven turns, bringing the working end beneath the turns, and up through the eye of the standing end. (Fig. 34).

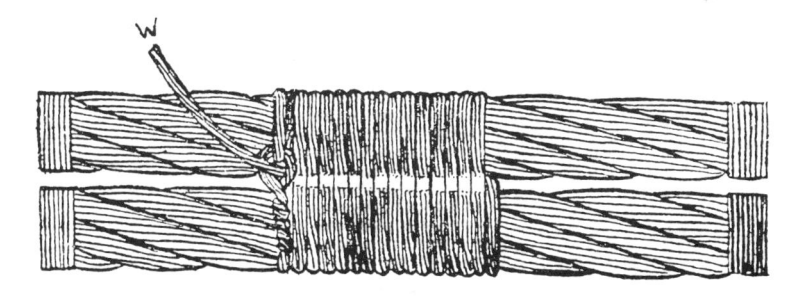

Fig. 34 — The Round Seizing.

36 MARLINE-SPIKE SEAMANSHIP

Now with the mallet commencing from the eye, haul each turn taut, to bite into the yarn serving, hauling taut the slack through the eye, then pass a second row of ten turns hand taut, bringing the working end up on the inside of the last turn, (Fig. 35), (I find that this little dodge of inside the last turn keeps the last turn of the upper row taut, during the passing of the frapping turns), and heave a steady taut. Always when seizing have

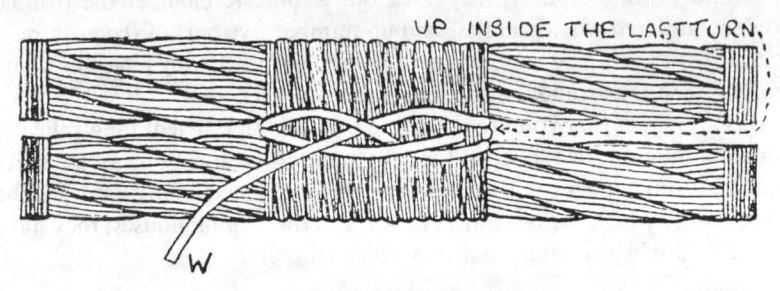

Fig. 35 — The Round Seizing. Stage 2.

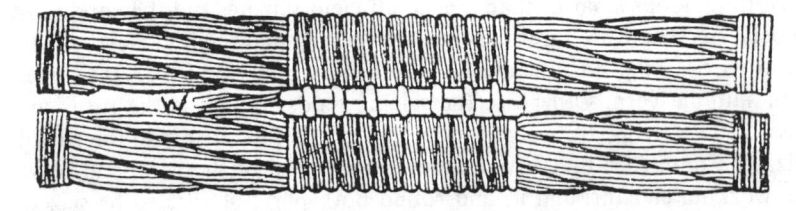

Fig. 36 — The Round Seizing. Stage 3.

one more turn on the bottom row than the top, for if you take an equal number of turns top and bottom, well it is obvious that the outer turn at each end of the top row will slip down. Complete the seizing by passing the frapping turns (which closely resemble the clove hitch), working each part taut separately, expending the end round and round the frapping turns on either side. This gives strength to the turns, also a fancy appearance. (Fig. 36). The round seizing is used when the strain is equal on both parts.

The Racking Seizing — Begin in the same way as for round seizing, but instead of taking turns round both parts of wire, dip or weave the seizing wire between the two parts in a figure of eight fashion to form a racking, heaving them taut as you go along, to say thirteen turns, then form the second row of turns by laying the first turn inside the last turn of the racking and work towards the eye of the seizing, the round turns lying in each case between the racking turns, finish off in the same way as for round seizing. The racking seizing is used when the strain is unequal on both parts. (*See* Fig. 37).

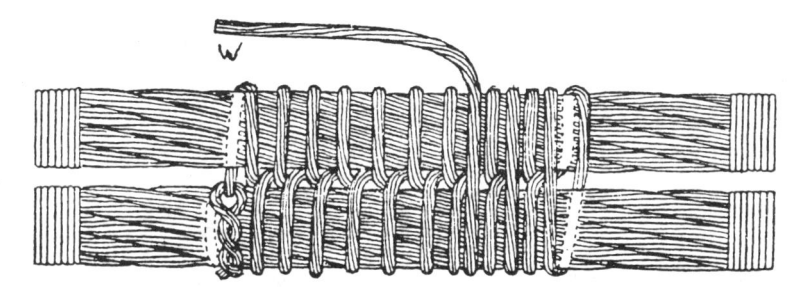

Fig. 37 — The Racking Seizing. Twist or Splice.

The Flat Seizing — The flat seizing is the first half of the round seizing, then pass frapping turns to complete. It is a light seizing.

On Span Making

A span consists of a wire grommet or a short splice strop, into which is inserted in either end a thimble. Spans are usually shackled to blocks to give the correct lead when hoisting boats etc., (Fig. 38), shows a grommet on the rigging block with thimbles inserted, ready for seizing in. The principal thing in span making is to keep the grommet as taut as possible whilst seizing in the thimbles. (Fig. 38), shows the windlass method which

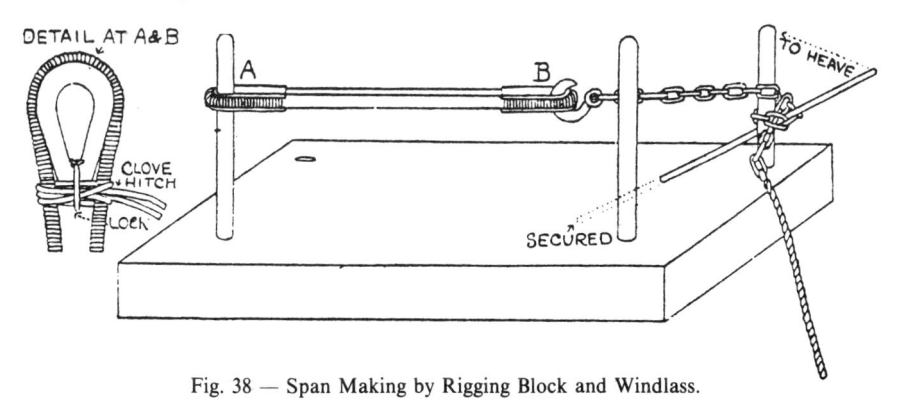

Fig. 38 — Span Making by Rigging Block and Windlass.

is to my mind the best; another method is by using a tackle either end. With the span on the stretch, alternately heave on the clove hitch at either end with the mallet.

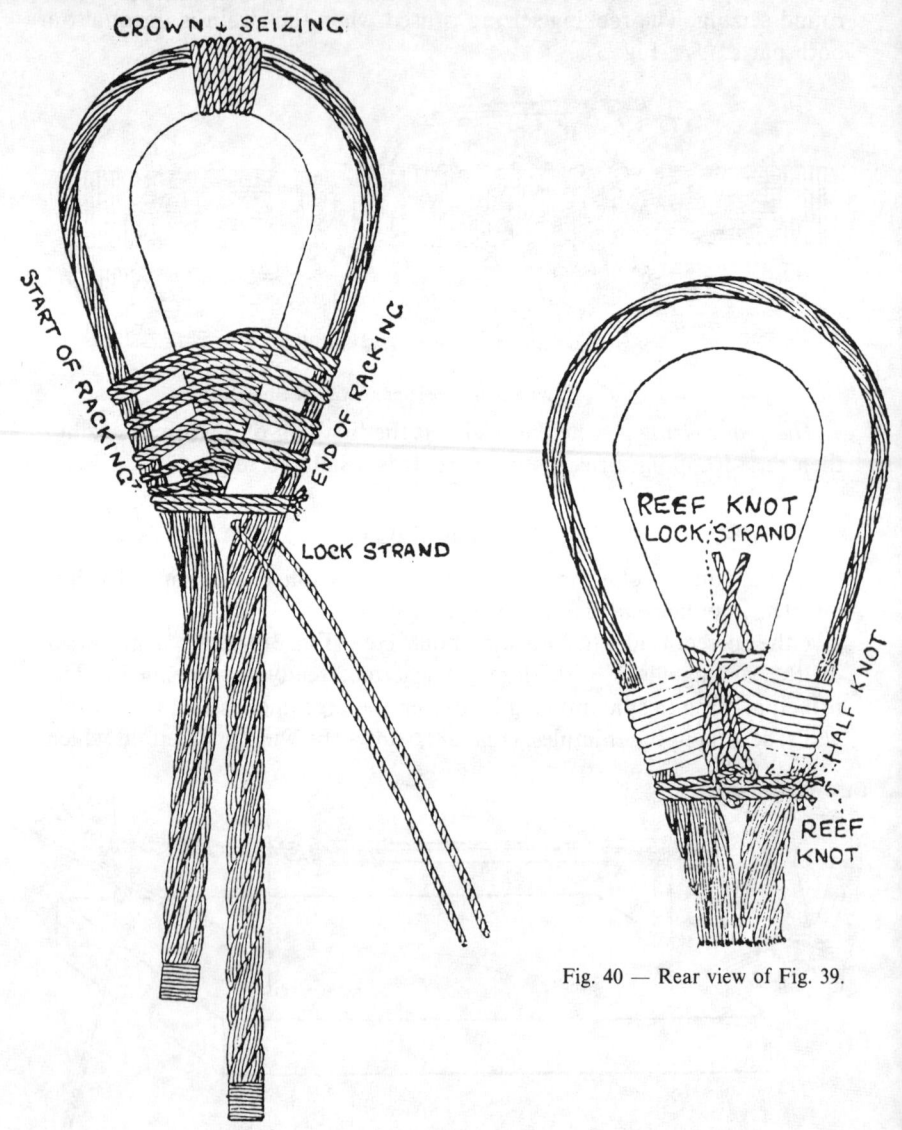

Fig. 39 — Seizing in the smaller Thimble with racking passed loosely. Vice omitted.

Fig. 40 — Rear view of Fig. 39.

MARLINE-SPIKE SEAMANSHIP 39

These hitches are prevented from slipping down by locking strands being taken round the hitches, and knotted on the inside of the toe of the thimble. It will be noticed that when heaving taut the hitches, the span tends to take half a turn one way or the other. This can be avoided by inserting a spike or iron bar in the thimble, and levering and keeping out the half turn as you heave the mallet. The thimble being seized in, complete by passing a round seizing, close up to each thimble. Commencing away from and working towards the thimble. Then remove the clove hitch. In a long span apply a flat seizing midway between the two round seizings.

On Seizing in Small Thimbles

Thimbles can be seized in by way of the heaving mallet and yarns, rigging screws, or the vice. I prefer the vice myself. Commence by securing the thimble in the vice, crown uppermost, and having applied the choke serving, place the wire in the score of the thimble, not forgetting that the bottom of the choke serving should be exactly in line with the toe of thimble when finally screwed up. Commencing first with a clove hitch (but work the cross of the hitch to the inside of the thimble which brings the two ends upwards), make taut by the mallet, pass a good seizing at the crown. Now place both parts of wire in the vice and screw up until the wire is hard up against the complete perimeter of the thimble, but don't crush the wires, then passing a seizing on each shoulder to prevent the wires at that point from lifting during the act of splicing. Then take the locking yarn and reeve it through the aperture between toe of thimble and point of splicing. Now take a long yarn, middle it, and reeving the ends through the bight, pass the racking, commencing from as low down as possible, working up the thimble as for racking seizing.

When enough turns have been taken, separate the two ends and half knot them in rear of the thimble at the toe. Now with the two ends, pass two turns round both wires, just below toe of thimble, finishing off with a reef knot. Then bring up both ends of locking yarn, haul taut and reef knot to finish on top of racking (*See* Figs. 39 and 40). This method is ideal for small wires, and if the racking is passed efficiently the choke serving can be dispensed with. It will be noticed that the two turns passed immediately below the toe prevents the loss of that extra bit of squeeze usually lost on unscrewing the vice. I find it advisable not to whip and prepare the strands before seizing the wire in, for with the wire in a virgin state, a better performance is made of the seizing in. Always stretch the seizing yarns and lightly grease them before using, they render better.

The Cropper or Gag Method of Securing Large Thimbles

This method is generally used when securing the larger wires to thimbles. Commence as in the other method described and having got sufficient squeeze with the vice, take two long yarns for the toe seizing and reeve then through the thimble, allowing them to lie on the inside of the toe, the ends hanging down and of even length. Now take about 30·5 centimetres of 10 millimetres circumference seizing wire and take one complete round turn, immediately below the toe of the thimble, and over the two yarns. Then pull or heave together both ends of the seizing wire with the pliers and twist them up (*See* Fig. 41). I usually take a turn with the mallet as for heaving, and using the vice as a leverage, pull gently taut and twist up at the same time. When the round turn is taut, cut off, leaving an end of about 51 millimetres and force the ends against the choke serving.

Now pass the two toe seizings, by pulling the two ends, first down then up taught, crossing them both first on the inside of the thimble, half knotting on the inside and outside of the thimble and hauling each half knot as taut as possible, as in West Country whipping; then a reef knot to finish (the twisted ends of seizing wire lying beneath the turns). Then repeat the operation with the two opposite ends. Thus, the seizing wire does the same duty as the two turns of yarn, as shewn in (Fig. 40). Always examine thimbles, some are crudely made and not symmetrical.

Thimbles Continued — It is the usual practice to serve over the wire that bends to the perimeter of the thimble. The theory is, that wire subject to such a sharp contour must lose its true lay or roundness, and with it some of its strength. There may, or may not be something in the theory, but the serving positively does preserve the wire. You must make sure that the wire does find a proper seating in the score, and does not ride on the rims of the thimble.

Thimbles are manufactured in various metals, to take wire of all sizes, which means that you must be reasonable, and not apply a serving of yarn too stout. One sees, from time to time, the other extreme: — A wire served over with an extra stout yarn, in an endeavour to make it fit a thimble, intended for a much larger wire. It matters little perhaps in

MARLINE-SPIKE SEAMANSHIP

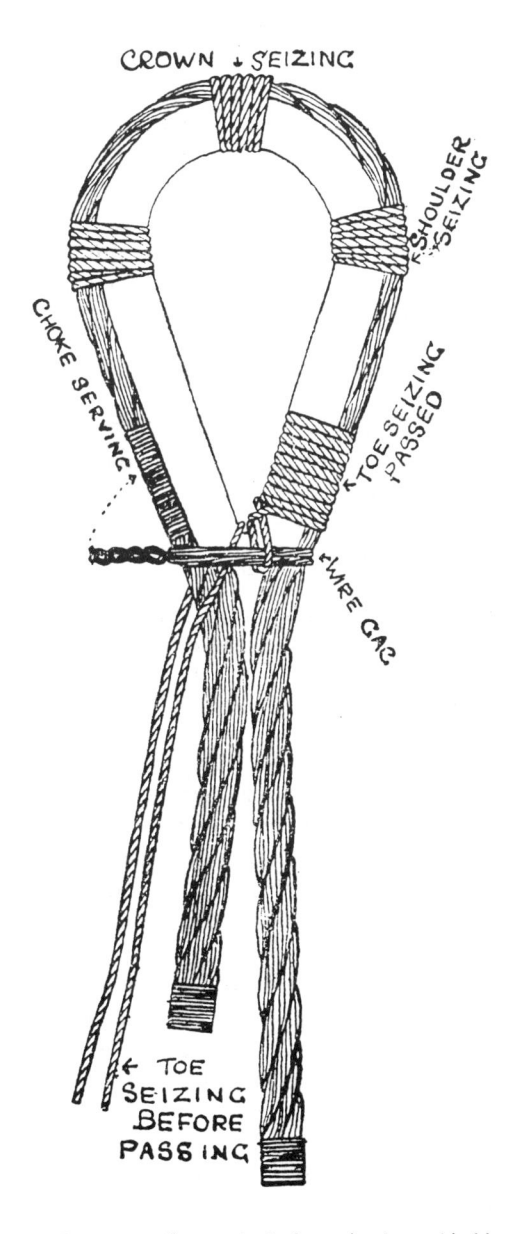

Fig. 41 — The Cropper or Gag method of securing large thimbles. Vice omitted.

standing rigging, but when the eye has to continually bear a great strain, or sudden jerks, the servings becomes compressed, then chafed, and finally parted from the wire. The result being a loose thimble.

Sometimes a brass or soft metal heart-shaped thimble comes our way, and a smart job can be made of the eye to be fitted, by first placing the thimble in the vice and with the hammer, open more the score at the points. You can then get your first tucks closer to, and under the points. On the completion of the splice, tap both points down to the splice, the result being, an extra taut thimble eye, with no visible space between the points of the thimble and the first row of tucks (Fig. 42).

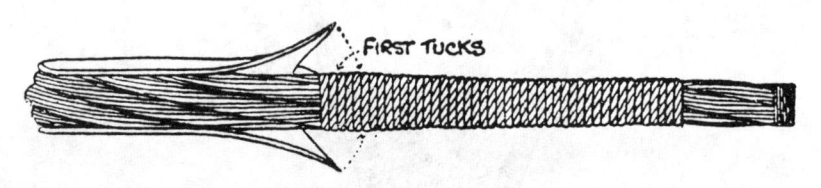

Fig. 42 — Widening the score of the Thimble at the points.

On Stropping Blocks

Since the introduction of the internal iron bound block, the stropped block has gone out of favour, but they are still in use. Blocks are stropped with a wire grommet of four parts. To find length of strand required. Take a piece of line or yarn about the size of the strand you intend using which should be preferably of flexible steel wire and take four turns around the block, in its score, the same round the thimble, and four times round the wire rope from which the strand came, and allow about 102 millimetres for the worming, parcelling, and serving of the grommet (Fig. 43).

It is amazing how this additional girth tends to shorten the circumference of the grommet. The eagle eyed ones can take the strand, lay up a few turns, place it in the score of the block and squeezing it with the fingers at the position for round seizing, can gauge the correct length, but only lots of practice will give you this skill. The grommet being completed, place it in the score of the block, taking care that you have the block the right way up, and with the splice of the grommet covering the square head of the pin of the block. The splice being the thickest part prevents the pin from withdrawing. Now seize the grommet to the tail of the block with two good yarns between the sheaves, and swallows.

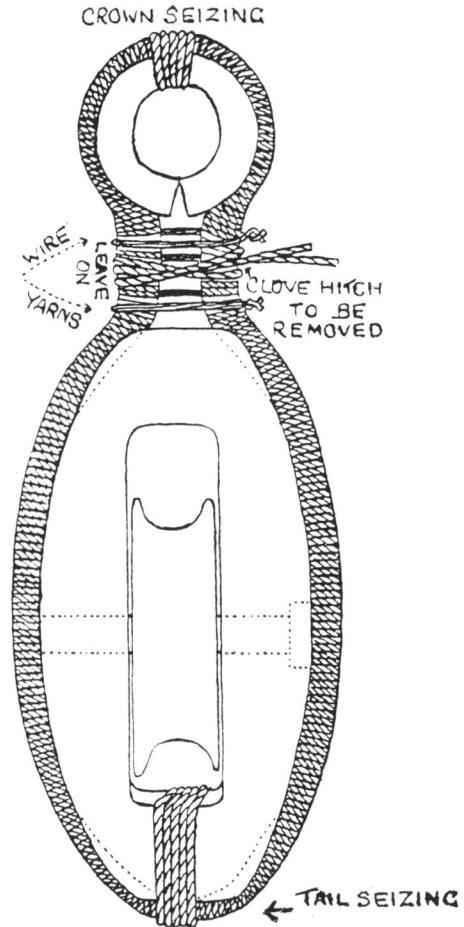

Fig. 43 — Stropping Blocks.

Then in the top of the grommet seize in the round thimble at the crown, then pass a clove hitch of stout yarn, between the thimble and the crown of block, and heave it as taut as possible with a mallet, and secure it. Now tap the grommet round the block, to its seating in the score. (*See* Fig. 43). Next take two pieces of light seizing wire, take a round turn on both parts of the grommet, one above, and one below the clove hitch, twisting them up as for gag method of seizing in a thimble. Now remove the clove hitch, and pass a round seizing of wire at the throat, covering up the thin wires. Blocks may be stropped by a length of wire short spliced, but this method is by no means as neat as the grommet. A disadvantage of the stropped block, is that to strip the block down for the cleaning of the pin and sheave, you must remove the strop. I like to see wooden blocks cleaned and varnished.

To Seize a Cleat on Davit or Stanchion

Most cleats today are of the iron variety, being either welded or rivetted to the position intended, but we still have the wooden cleat with us, and a nice job of work it looks when stained, and the seizings painted your favourite colour. Commence by stopping the cleat on at the required position, with a light seizing of yarn or thin wire, which can be

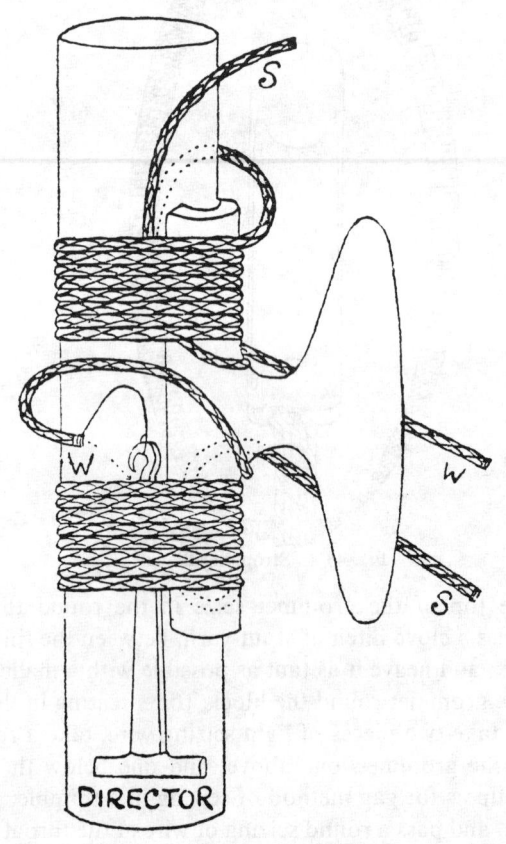

Fig. 44 — Fixing the Cleat.

left on and covered by the seizing wire, or if you are re-seizing a cleat, remove one old seizing at a time, and renew. (Fig. 44) shows standing part laid on davit, the working part passed round and round the davit, on the top of standing part to the required number of turns.

Heave each turn as taut as possible, then the feeler which resembles the old fashioned button hook, is force up, between the davit and the turns, and a strong thin yarn is secured to the working part of the wire, the other end of the yarn is secured to the hook of the feeler, the yarn and the next wire being drawn down between the turns and davit. Then heave the working end down and back as taut as possible, jamming it under the turns.

Now turn to the standing end which only requires hauling taut as for working end, then cut off short both ends with hammer and chisel. (*See* Fig. 44). The turns can be hauled taut with the mallet, but I prefer the following method, but it is a two man job: — if the davit is near the ship's side, and they mostly are, as I pass the turns, I lead the working end round a guard rail stanchion, my assistant holding the end firmly with one hand, lightly taps down on the wire between the davit and stanchion, with the mallet, at the same time taking in the slack of the wire, he then gives me the end back and I take another turn and he repeats the process. This method avoids kinking the wire, which occurs when using the heaving turn on the mallet.

Wire Net Making

These nets are generally used for hoisting in provisions and as access to boat nets. They are easy but very monotonous to make. It is in fact a job of all seizings. Commence by rigging up a wooden frame, slightly larger than the net you intend to make, i.e. four lengths of wood, bolted or nailed together, next knock in stout nails at equal distance apart to the size of mesh required. Then arrange the main or perimeter wire which should be 13 millimetres larger in circumference than the cross members or mesh wires, form an eye in the four corners and seizing same with round seizings. Now haul the permimeter wire taut round the outside of the nails, marry the two and short splice. Now arrange the vertical wire up and down and round the nails loosely, then interlock the horizontal wires over and under the vertical. Now splice one end of the vertical wire into the perimeter wire (two and a half tucks are sufficient).

46 MARLINE-SPIKE SEAMANSHIP

Then commencing from the spliced end, haul the vertical taut round the nails and stop the end to the perimeter wire. Then repeat the operation with the horizontal wire. Complete by passing a seizing of unannealed wire round each crossing of wires and tuck remaining two ends of mesh wires into the perimeter wires and serve over. A useful size for provisioning from being 4 to 4·5 metres square. The mesh wires may be tucked through the perimeter wire where they meet (*See* Fig. 45). The start of the cross seizing is shown in (Fig. 45).

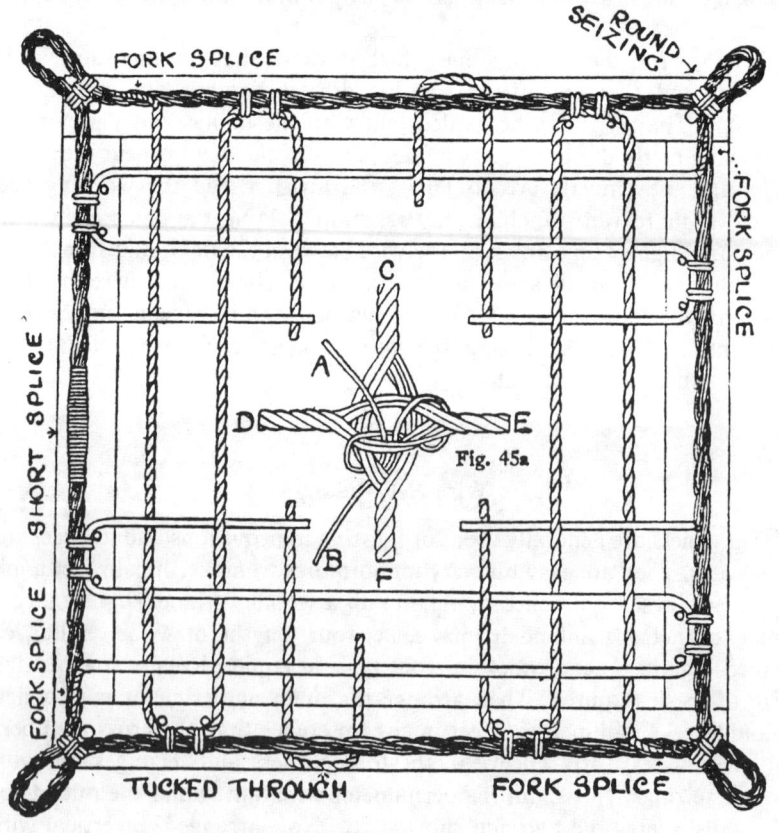

Fig. 45 — The Lifting Net of Wire.

The Cross Seizing in Net Making — Enter the spike through the centre of both wires at the crossing, and reeve the length of seizing wire through both parts, and middle same. Then first expend end *A* as taut as possible, between angles *C D* and *E F*. Next pass *B* between and angles *D F* and *C E*. Then twist *A* around *B* one full turn, and finish off, by taking three round turns with *A* and *B* on the lower wire, on either side of the crossing and then lifting one strand of the lower wire, take two round turns on it and cut off. (Fig. 45a).

On Making Jumping Ladders

Easy to make and useful to the yachtsman. Usually made of 13 millimetres circumference F.S.W.R. for the sides or stringers and greenheart rounds for the rungs or treads.

To Make a Ladder — Commence by middling the wire and inserting a suitable round thimble secured by a round seizing. Now get both wires on a stretch but not too taut, and proceed to enter the rungs at equal distances apart. This is where our old friend the pilot spike plays a part. Enter the spike under the strands, now place one end of the rung in the hollowed end of the spike, and force it completely through the wire with the rung. It will be found that the rung finds its own seating, the three wires on either side slipping into the rounded score of the rung. Now enter the spike in the opposite wire and repeat the operation, working along the two wires and towards the top end of the ladder, the top rung being secured as follows — a soft eye is fitted in each end of the two wires, which, when completed, will take the heart shaped thimble, the round seizing and the rung.

It is important that the two eyes are of exactly the same size, otherwise, when hung, the ladder will not be perpendicular. The splices of the two eyes being served over, the thimbles and the top rungs seized in, now apply a light serving of wire or yarn above and below each rung to keep them in position. If you do not happen to possess a pilot spike, enter two yarns through the centre of the wire, middle them and pull the wire apart. Insert the rung and withdraw the yarns. If you require a smart ladder, worm, parcel and serve over the stringers with yarn above and below each rung. Then cover up the serving with a strip of flax canvas of the required length and breadth. This is done as follows: — Measure off and cut out the holes to allow the ends of the rungs to protrude, then buttonhole stitch the holes to prevent them tearing, now place the holes over the ends of the rungs and stitch up on the inside of the stringers (*See*

48   MARLINE-SPIKE SEAMANSHIP

Fig. 46), and if you want the ladder extra smart, apply turks' heads above and below each rung on the stringers.

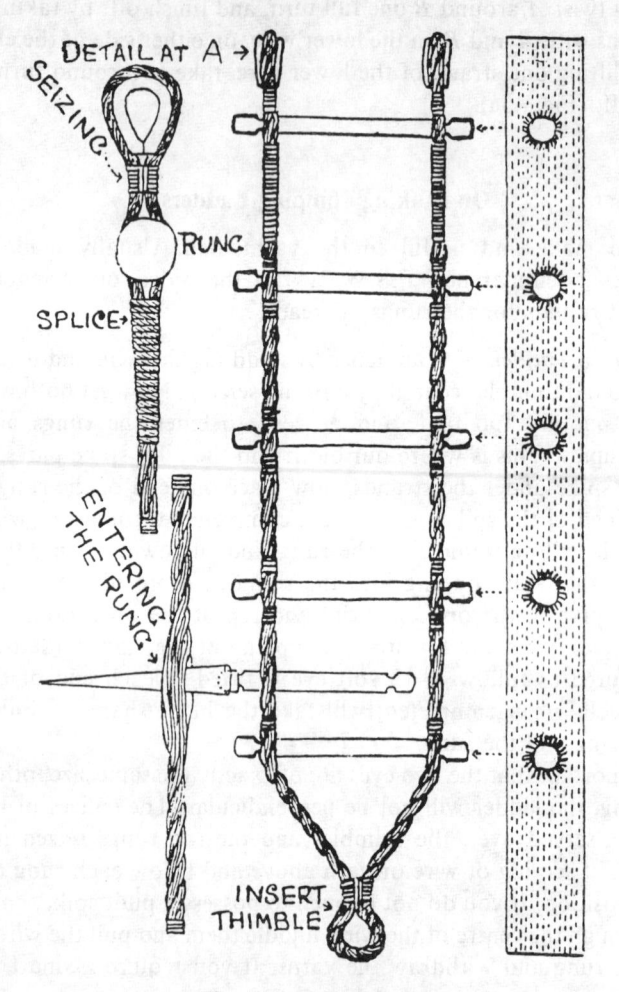

Fig. 46 — Ladder Making.

Fixing the Tulip

The metal tulip so called, because it resembles that flower in shape, is fitted in the end of the wire guard rails of ships. Its advantage over the guard rail fitted with an eye either end, is that it can be unrove from the guard rail stanchions when necessary. Commence by serving the end for fixing with yarn, or thin seizing wire about 51 millimetres longer than the petals, then enter the end inside the petals, and hard up to the eye. Now place wire and tulip in the vice and screw up taut. Then with a sharp and slender pricker, pierce the wire through the holes in the tulip from either side, taking care that you obtain a clear passage for the entry of the rivets. Copper nails of the right diameter are ideal. Enter them with the heads the reverse of each other, then cut off and rivet up. Keep wire and tulip taut in the vice, when piercing and entering rivets. (*See* Fig. 47).

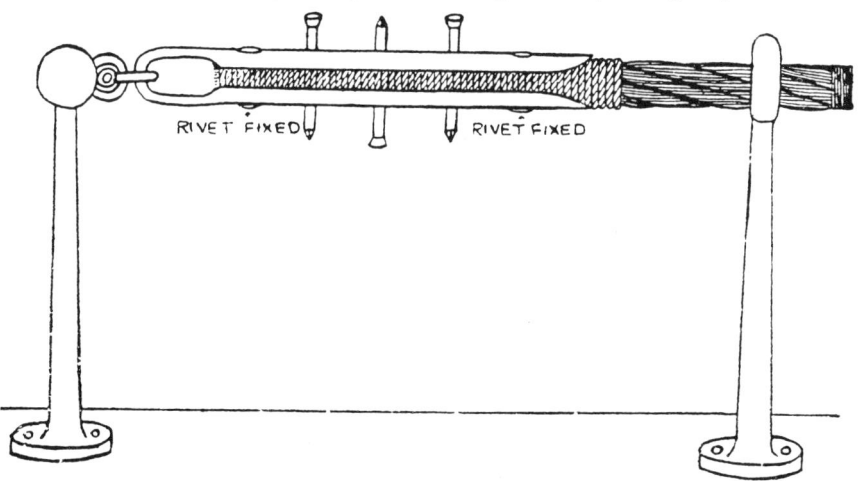

Fig. 47 — Fixing the Tulip.

The Bulldog Grip or Clip

A very useful gadget to have, it is made to take wire of all sizes and it can be used for seizing in thimbles. Here again a little trick is required to keep it in position whilst screwing up the two nuts. After securing the grip on both parts of the wire loosely, and close up to the toe of the thimble, pass a length of seizing wire or yarn round both parts of the "U" bolt, and in between the two wires, and reef knot on the inside of the toe of the thimble as for locking strand in span making, and alternately screw up the nuts a little at a time until you get a good compression of the wires when under full strain. When employing more than one bulldog grip, say for a makeshift eye, always put the base of the clip to the standing part, which should be the part bearing the most strain, spacing clips at intervals of five times the circumference of the wire. (*See* Fig. 48).

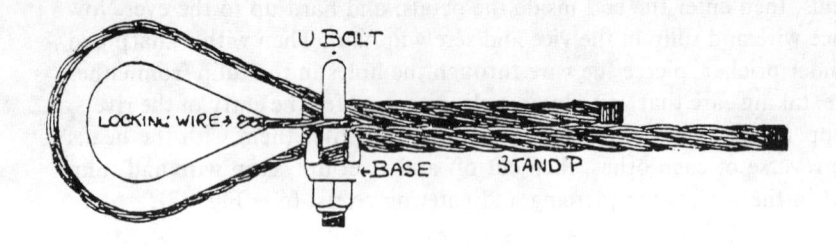

Fig. 48 — Fixing the Clip.

The Standing Turk's Head

Called the standing turk's head because it is unmovable on the wire rope when completed. It is made with two lengths of annealed wire, one long and one short. Commence by tucking the longer length under two strands of the wire, hauling it through and middling to make two ends. Then tuck the shorter length under two strands and bury its standing end into the wire, taking care that its entry into the wire is directly in line with the longer length. Now form a wall knot away to the left and haul it taut. Then form a crown knot to the right hauling it taut. Complete by following all parts round three times using the three ends, hauling the ends taut and cutting off to finish. This brand of turk's head head makes a useful distance mark on the fog buoy wire, or a collar for steadying lines of boat slings. (Fig. 49).

To Finish off an end of Wire not Required to be Spliced — It may be that after fitting an eye in one end of a wire, that you decide to leave the other end unfitted, but preserved, and be able to handle it without fear of tearing the hands. I use the following method: — Whip the end neatly 51 millimetres, with sail twine or the like. Then make a sleeve of flax canvas about 152 millimetres in length, and just a fraction larger in circumference than the wire, but instead of using canvas for the bottom of the sleeve use pig hide; this prevents the ends of the wire from piercing through the bottom of the sleeve. Now give the end a spot of grease and

MARLINE-SPIKE SEAMANSHIP 51

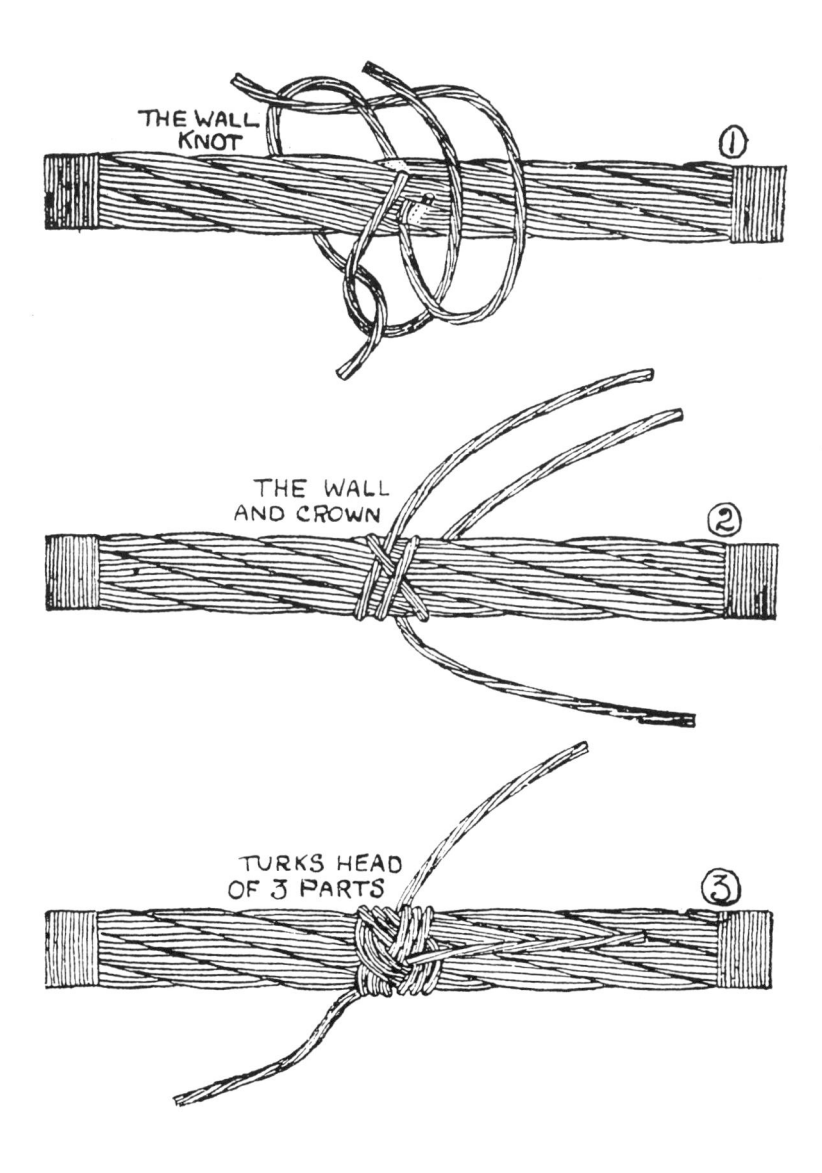

Fig. 49 — The Standing Turk's Head.

slip it into the sleeve hard on to the hide bottom. Complete by passing a serving of yarn over sleeve (*See* Fig. 50).

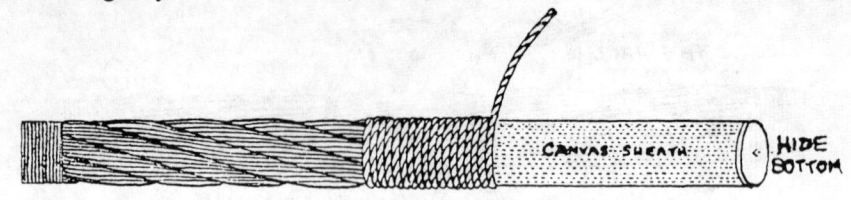

Fig. 50 — Preserving a Wire End.

On Making Wire Grommets

Wire grommets are easy to make, and they are made up in the following constructions. Three, four, five and six parted with the ends rolled into form the heart, known as the perfect wire grommet. They are used for various purposes, for strops on lower booms of ships, to which

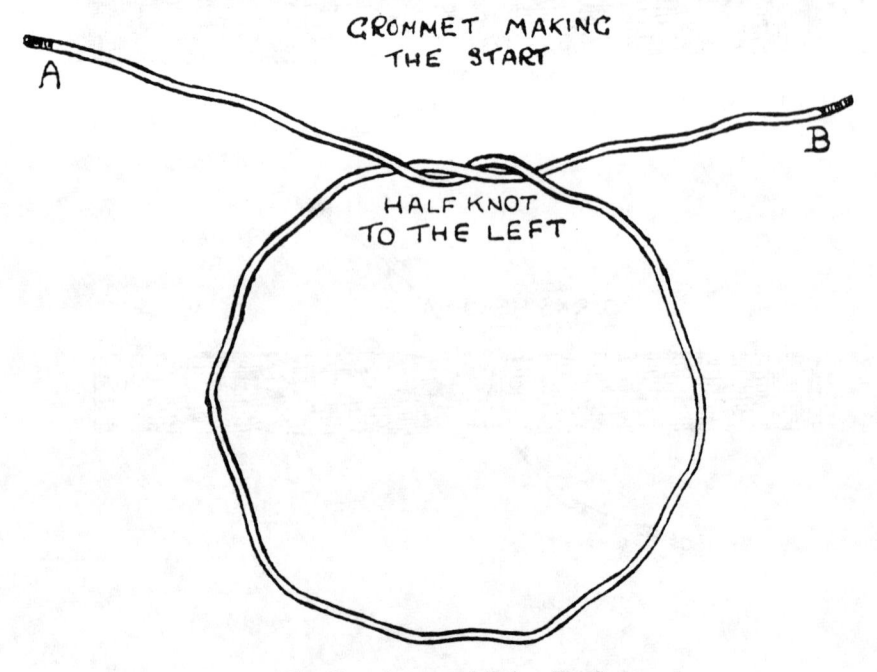

Fig. 51 — Grommet Making. The Start.

are shackled the ladders and lizards for securing boats, and strops for blocks. Three and four parted grommets also make light and handy strops when squeezed together and seized near the ends. Some argue that the grommet is expensive to make from a six stranded wire. If you only make one grommet it naturally would be, but if six grommets are made and needed there can be no loss in using a six stranded wire.

A Note on Stranding Wire Rope — I find from experience that the best way to overcome this tedious business is as follows:— Middle the rope and bend on at that position a line and trice it up until the two ends are about waist high. Then proceed to unlay all strands on one side of the hitch, cross over to the other side and unlay them as close to the hitch as possible.

You will find on lowering down the wire and removing the hitch that you will be able to shake the strands free, but don't forget to whip all the strands, even though you may only need the one strand.

How to Make — Commence as in (Fig. 51) by passing end B over and under each end A to the size of grommet required, then continue to pass B over and under A until we arrive at the second stage (Fig, 52), showing B end passing on the inside of A which gives us two parts. Then lay up a third part using end B which again must pass on the inside of A.

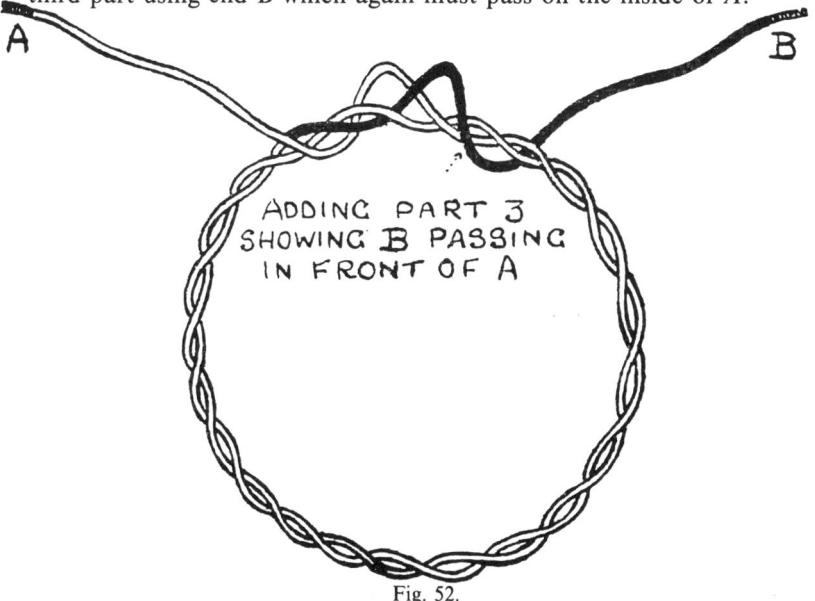

Fig. 52.

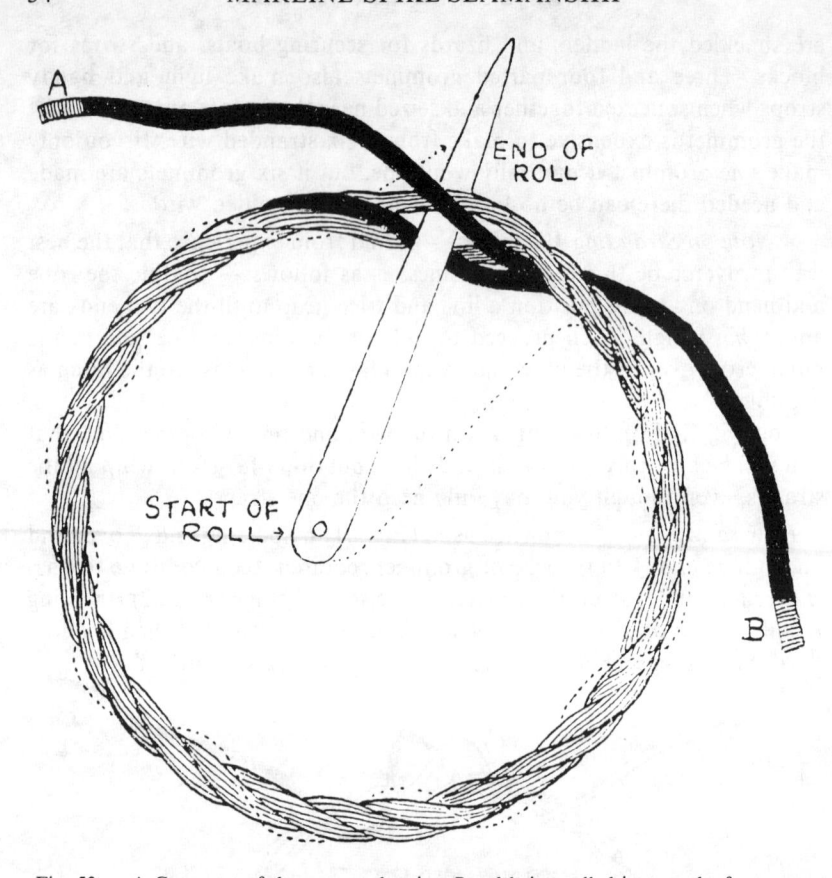

Fig. 53 — A Grommet of three parts showing *B* end being rolled in to make four parts.

Fig. 53a — Finishing a Gromet of four parts and above, also type "A" reduced Eye.

This is very important, for you should pass *B* end on the outside of *A* at this stage, and continue to the fourth and final stage you will find that when you commence to tuck and complete your four parted grommet that you have lost the true lay and that you are out of alignment for tucking. You may, if you wish to, pass *B* end on the outside of *A* at the second stage, but if you do so you must continue passing *B* on the outside of *A* as you add every extra part. Two, three and sometimes four parts can be laid up easily by hand, but I prefer to roll number four in with the spike. (*See* Fig. 53). Now complete the four parted grommet by tucking *B* end over *A* and under two (*See* Fig. 53a) strands of the grommet, then tuck *A* end over *B* and under two strands, now remove the hearts from the strands and repeat the operation. The second tucks each way being completed, lightly hammer the tucks down, then twist and break off the yarns.

To make a five parted grommet lightly tap another part in. I can easily manage to roll number five part in with the spike about 76 millimetres at a time, applying a light seizing at every 76 millimetres to keep it in the lay, finishing off as for four parted grommet (Fig. 54). It will be noticed that in describing the construction of grommets that I used *B* as the working end throughout, but both ends may be utilised as working ends, keeping both ends the same length, when each additional part has been completed.

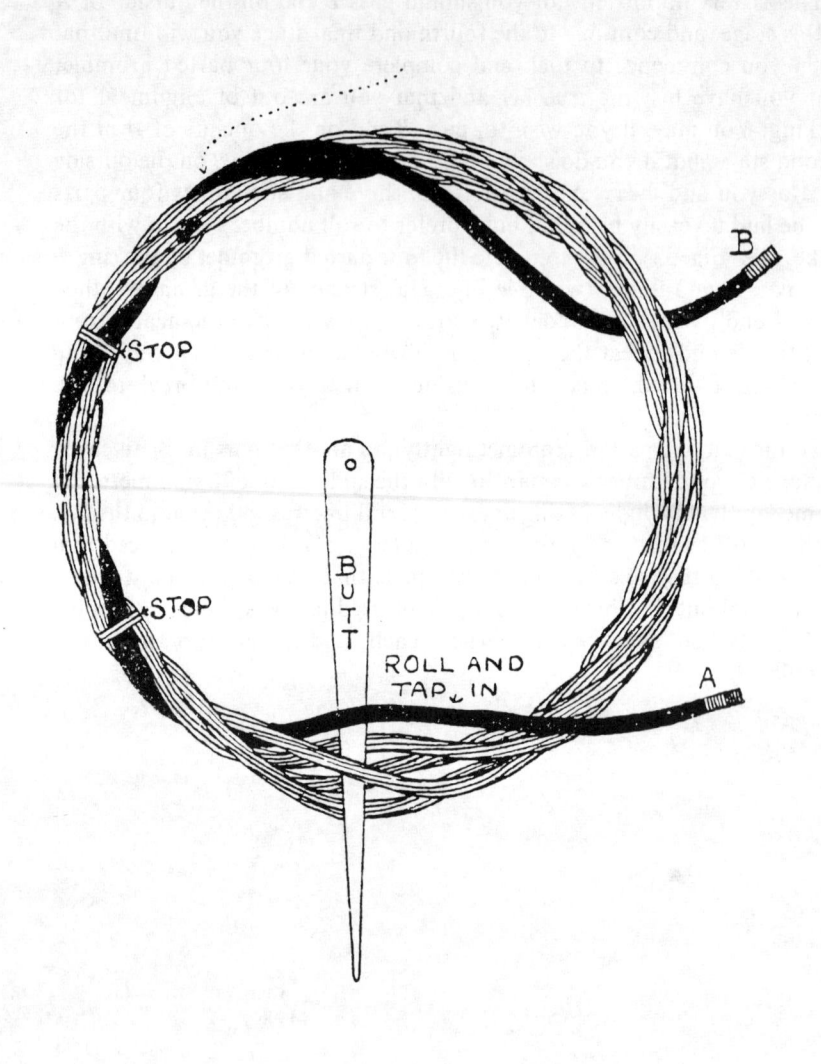

Fig. 54 — A Grommet of four parts laying up a fifth part using end *A*.

The Perfect Wire Grommet

This is a work of art. The craftiest job ever with a wire strand. Let us assume that you have laid up five parts of the grommet and applied a seizing round *A* and *B* ends at the junction. Now take the spike and lift up two strands of the five and enter one end of what I call the pilot strand

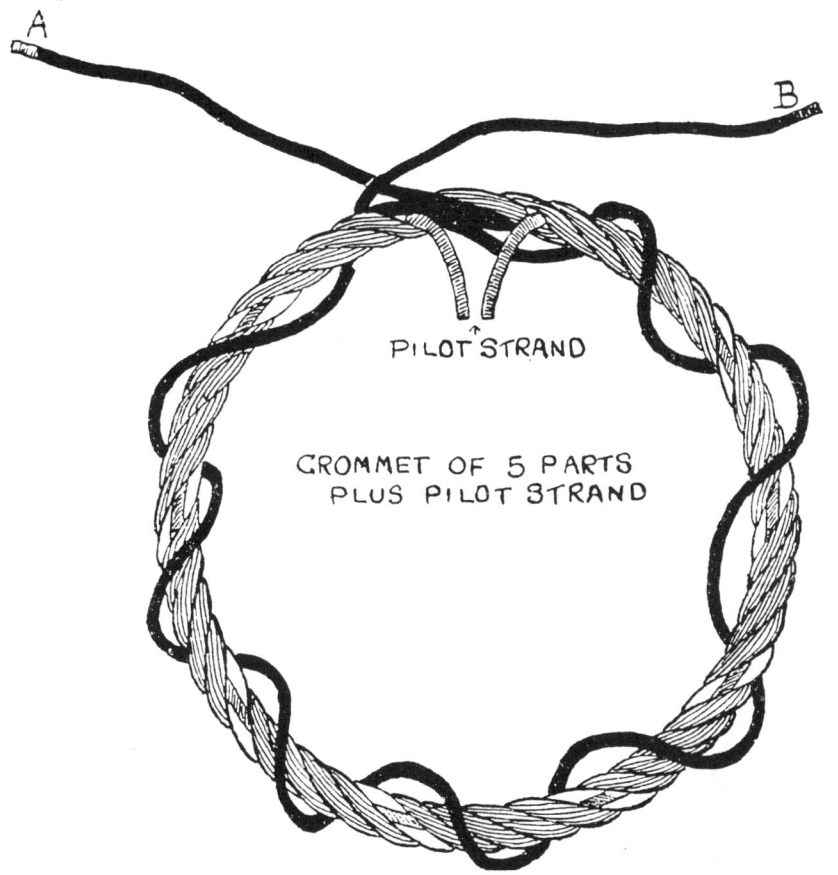

Fig. 55 — The Perfect Wire Grommet. Adding part six using end *B*.

against the lay, allowing about 76 millimetres to protrude (this strand being of the same construction and size of the strand with which you are making the grommet and about 152 millimetres longer than the circumference). Then entering the spike over the pilot strand and under the next two strands immediately above it, roll the pilot strand in to form a temporary heart rolling the spike towards the right as in the four parted grommet, which when completed should appear as in (Fig. 55). It will now be found that a sixth part can be worked in easily by hand, and it should cover up the pilot strand. The sixth part having been laid in, apply a light seizing at the junction. Now tuck *A* and *B* ends through the centre of the grommet, which gives you three parts of the grommet on either side of ends *A* and *B* and they should be pointing in opposite directions, when pulled close in to the lay of the grommet (Fig. 56).

Next comes the operation of rolling *A* and *B* ends into the grommet to form the permanent heart, and at the same time of removing the pilot strand. This is done as follows. Place the spike over the end you intend to roll in, and under two strands of the grommet, then pull out about 76

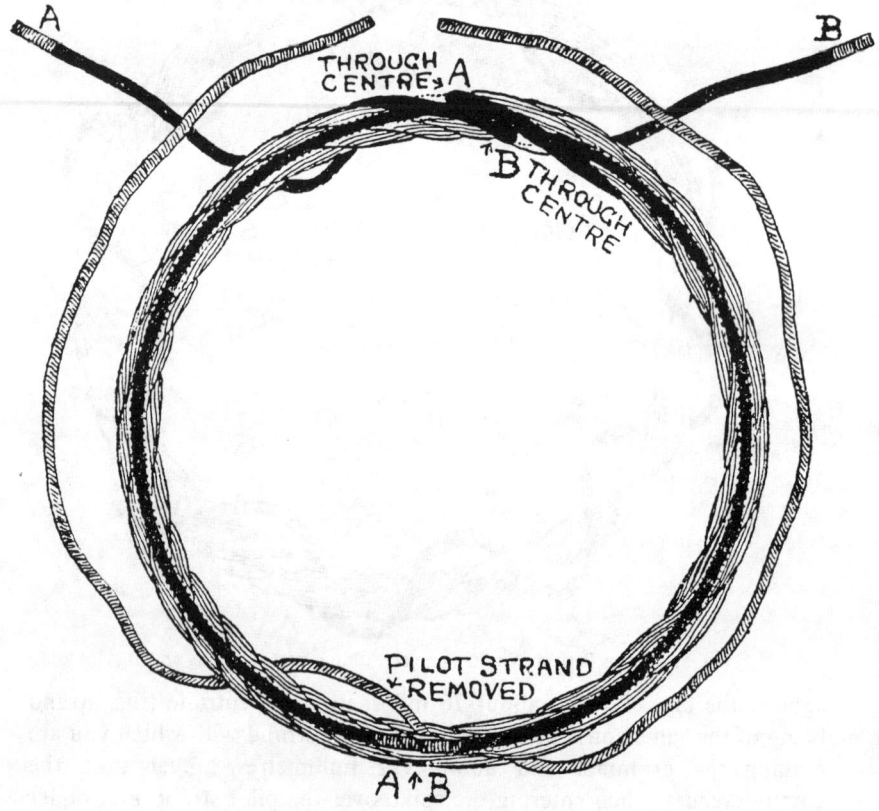

Fig. 56 — Final Stage. Rolling Home *A* and *B*.

millimetres of the pilot strand, now roll in the end of the grommet that distance, repeating the operation to about half the circumference of the grommet, then turn to the other end and repeat the operation. You will now be able to remove the pilot strand from the grommet, completing the grommet by burying the two ends, which should meet close up to each other on the inside of the grommet, both ends being neatly whipped (Fig. 56).

You may or may not be successful in making the perfect grommet at the first attempt, but when success is achieved you will have been amply repaid for your patience and perseverance. See to it that you have no turns or kinks in the strand when you commence to lay up, that you find the correct and true lay as you proceed to add the extra parts.

Another Method of Grommet Making — A rather neat grommet of four parts can be made as follows:—

Carefully unlay from the wire two strands together in their true lay, instead of one. The two strands to be used should be about three and a half times the circumference of the grommet required. Commence as shown in (Fig. 57), laying up *A* and *B* until the four ends oppose each other at the juncture. The four ends are then interlocked and correctly mated (*See* Fig. 58). Then one strand is unlayed and the vacancy in the grommet being filled by the opposing mated strand, which should

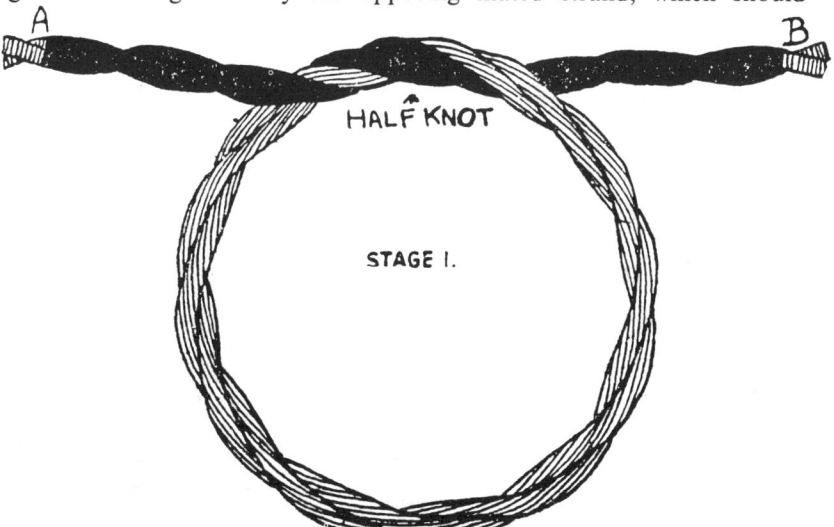

Fig. 57 — The Four Parted Grommet using the Strands.

terminate at the position shown in (Fig. 59). The four ends may then be tucked as already described, or rolled into the heart of the grommet.

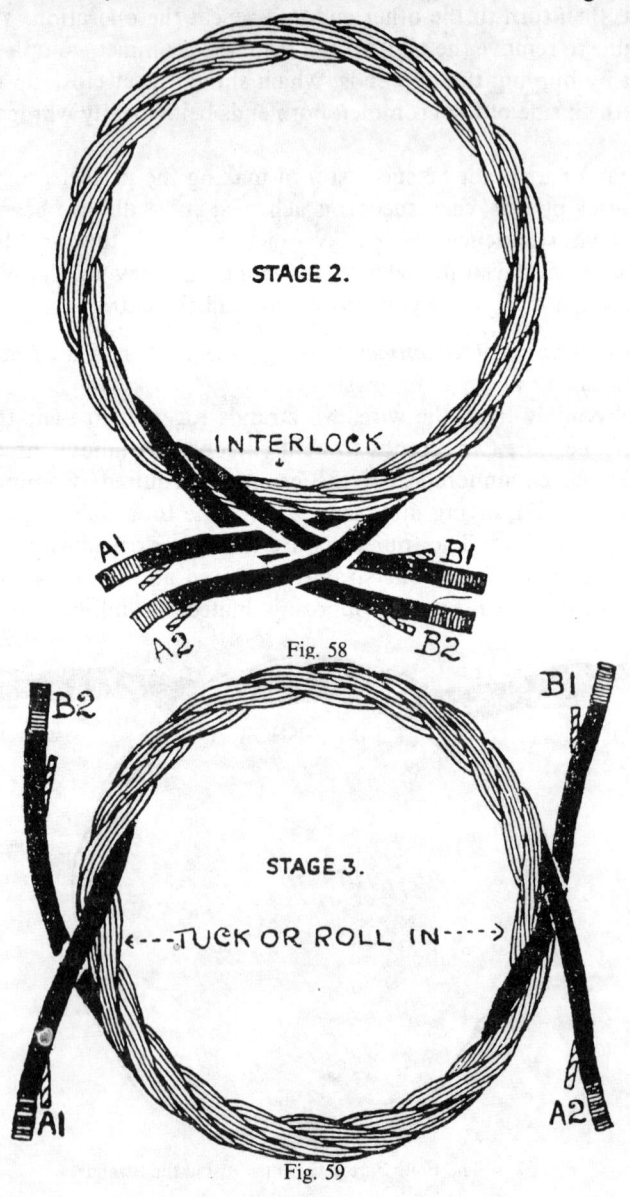

STAGE 2.

INTERLOCK

Fig. 58

STAGE 3.

<---TUCK OR ROLL IN--->

Fig. 59

MARLINE-SPIKE SEAMANSHIP

The Ribbon or Lifting Band

It is made of extra special wire of various sizes. It is the ideal strop for lifting heavy cylindrical weights, such as torpedoes, etc. It can be made any length or width. Supposing you want a band of 2·5 metres in length, with wires a width of six wires. You would of course, allowing for the

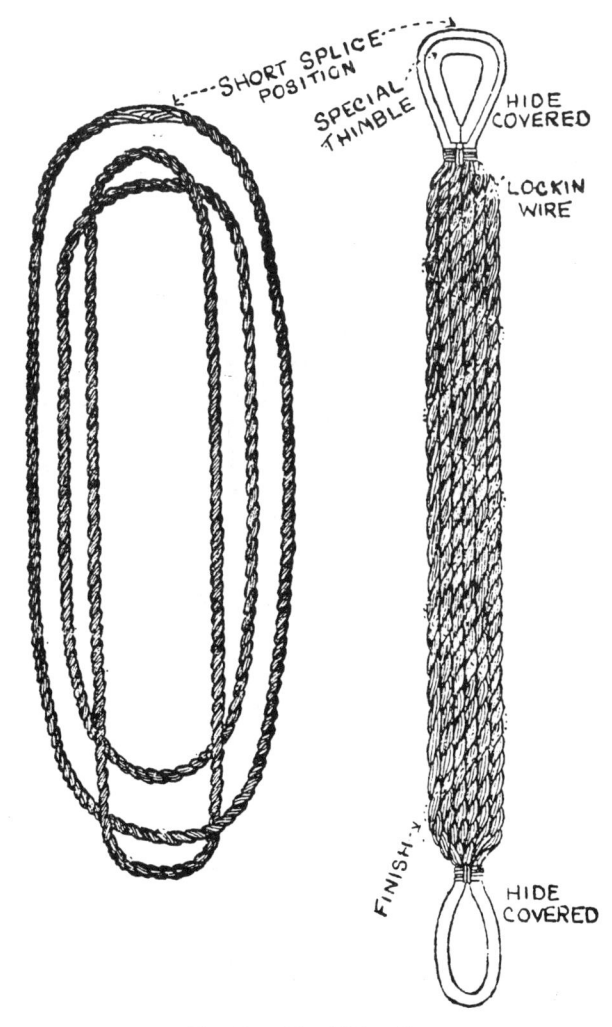

Fig. 60 — The Ribbon Strop.

short splice, need approximately 15·3 metres of wire. Commence by marrying the wire at 14·6 metres, and short splice the double up (Fig. 60), and stretch the loops out taut, which gives you a width of six wires, arranging the short splice to fit in the crown of the lanyard type of thimble. Next cover the splice and its accompanying wires with cow hide, to the perimeter of the thimble, the stitches being on the inside. Then seize in the thimble and apply a round seizing of 4 millimetres copper wire, taking care that the wires are of equal length and tautness. Now form the wires or loops of the opposite end, of which you are to form a soft eye, and cover with cow hide, the stitches in this case being on the outside of the eye, then apply the round seizing as for the thimble eye. Both eyes completed, the next and final operation is of locking the six wires together thus, stretch the band out taut by a tackle in each eye, then working from the thimble eye, with a long thin spike pierce all six wires through their centres diagonally, and pass through 4 millimetres circumference copper wire finishing on the outer wire close to the seizing of the soft eye, by taking two round turns on one strand of the wire and cut off (Fig. 60). To fix, reeve thimble eye through soft eye, and hook on.

Finally I hope that I have in some way added to the works of my predecessors in preserving this ancient art. I have endeavoured to keep rigidly to the practical side of rigging and expound it in the language of the rigger, but it must be realised that this little book does not contain the full treatise of the art. The tables of breaking and working loads, construction and the care and maintenance of wire etc., have been admirably described in other works.

HANDY BOOKS
on
ROPE, WIRE and CANVAS

Harrison Book of Knots	By Capt. P.P.O. HARRISON
Knots, Splices and Fancy Work	By CHARLES L. SPENCER
Advanced Rope Working	By LEONARD POPPLE
Rope Splicing	By P. W. BLANDFORD
Wire Splicing	By R. SCOT SKIRVING
Cordage and Cables	By Capt. P. J. STOPFORD
Netmaking	By P. W. BLANDFORD
Working in Canvas	By P. W. BLANDFORD

Prices on Application

BROWN, SON & FERGUSON, LTD.
4-10 DARNLEY STREET, GLASGOW, G41 2SD